Guy Hinsdale

Acromegaly

Guy Hinsdale

Acromegaly

ISBN/EAN: 9783337379827

Printed in Europe, USA, Canada, Australia, Japan

Cover: Foto ©berggeist007 / pixelio.de

More available books at **www.hansebooks.com**

AN ESSAY TO WHICH WAS AWARDED THE BOYLSTON
PRIZE OF HARVARD UNIVERSITY FOR
THE YEAR 1898

BY

GUY HINSDALE, A.M., M.D.,

Fellow of the College of Physicians of Philadelphia and of the American Academy
of Medicine; Member of the American Neurological Association and American Climatological Association; Assistant Physician to the Orthopedic Hospital and Infirmary for Nervous Diseases, and
to the Presbyterian Hospital in Philadelphia, etc.

REPRINTED FROM
MEDICINE, 1898
WILLIAM M. WARREN, PUBLISHER
DETROIT

Nulla autem est alia pro certo noscendi
via, nisi quamplurimas et morborum et
dissectionum historias, tum aliorum tum
proprias collectas habere, et inter se com-
parare.

MORGAGNI, *De Sed. et Caus. Morb.*

BOYLSTON MEDICAL PRIZES.

These prizes, which are *open to public competition*, are offered annually for the best dissertations on questions in medical science proposed by the Boylston Medical Committee.

At the annual meeting in Boston in 1898, a prize was awarded to Guy Hinsdale, M.D., of Philadelphia, Penn., for an essay on *Acromegaly*.

For 1899 two prizes are offered : —

1. A prize of one hundred and fifty dollars for the best dissertation on *The results of Original Work in Anatomy, Physiology, or Pathology*. The subject to be chosen by the writer.

2. A prize of one hundred and fifty dollars for the best dissertation on *The Results of Original Investigations in the Psychology of Mental Disease*.

Dissertations on these subjects must be sent post-paid to W. F. WHITNEY, M.D., Harvard Medical School, Boston, Mass., on or before *January 1, 1899*.

For 1900 two prizes are offered : —

1. A prize of one hundred and fifty dollars for the best dissertation on *The results of Original Work in Anatomy, Physiology or Pathology*. The subject to be chosen by the writer.

2. A prize of one hundred and fifty dollars for the best dissertation on *The method of Origin of Serpentine Arteries and the Structural Changes to be found in them. Their Relation to Arterio-capillary Fibrosis, Obliterating Endarteritis and to Endarteritis Deformans*.

Dissertations on these subjects must be sent to the same address as above on or before *January 1, 1900*.

In awarding these prizes preference will be given to dissertations which exhibit original work, but if no dissertation is considered worthy of a prize, the award may be withheld.

Each dissertation must bear in place of its author's name some sentence or device, and must be accompanied by a sealed packet bearing the same sentence or device, and containing within the author's name and residence. *Any clew by which the authorship of a dissertation is made known to the committee will debar such dissertation from competition.*

Dissertations must be written in a distinct and plain hand, and their pages must be bound in book form.

All unsuccessful dissertations are deposited with the Secretary, from whom they may be obtained, with the sealed packet unopened, if called for within one year after they have been received.

By an order adopted in 1826, the Secretary was directed to publish annually the following votes : —

1. That the Board do not consider themselves as approving the doctrines contained in any of the dissertations to which premiums may be adjudged.

2. That in case of publication of a successful dissertation, the author be considered as bound to print the above vote in connection therewith.

The Boylston Medical Committee is appointed by the President and Fellows, and consists of the following physicians : ROBERT T. EDES, M.D., *President;* WILLIAM F. WHITNEY, M.D., *Secretary;* H. P. BOWDITCH, M.D., FRANK W. DRAPER, M.D., J. COLLINS WARREN, M.D., SAMUEL G. WEBBER, M.D., F. H. WILLIAMS, M.D., EDWARD S. WOOD, M.D.

The address of the *Secretary* of the Boylston Medical Committee is WILLIAM F. WHITNEY, M.D., Harvard Medical School, Boston, Mass.

TABLE OF CONTENTS

ACROMEGALY: AN ESSAY TO WHICH WAS AWARDED THE BOYLSTON PRIZE OF HARVARD UNIVERSITY FOR THE YEAR 1898.

BY GUY HINSDALE, A.M., M.D.,

Fellow of the College of Physicians of Philadelphia and of the American Academy of Medicine; Member of the American Neurological Association and American Climatological Association; Assistant Physician to the Orthopedic Hospital and Infirmary for Nervous Diseases, and to the Presbyterian Hospital in Philadelphia, etc.

It was reserved for M. Pierre Marie, about ten years ago, to recognize the identity of this disease. Isolated cases of this affection had, no doubt, attracted notice all over the world, but they were regarded as peculiarities of an individual, as an example, perhaps, of local hypertrophy; or, as in special cases warranting clinical description, they were designated under the names of macrosomia, general progressive hypertrophy, prosopectasia, general hyperostosis, ankylosis of vertebral column with deformation of bone; or, in certain cases, were generally recognized as giants. Nevertheless, in 1886, M. Marie, who was at that time the director of the laboratory in La Salpêtrière, noticed the similarity of two patients, at that time in the wards. A study of these cases convinced him that it was not an accidental resemblance, but the outgrowth of a pathologic change, depending upon some alteration in the central nervous system. He was able to connect the manifestations in life with a hypertrophy of the pituitary body, and thus gave to medical nomenclature a new term, Acromegaly, or, as many chose to call it in honor of the observer, "Marie's Malady." A French writer calls Marie "le père d'acromégalie."

"Pachyacria" has been proposed by Mosler and Arnold. Prof. Cunningham, and Haughton and Ingram, of Dublin, proposed the name "megalacria," a term rationally constructed and at the same time euphonious. It is probable, however, that in this case, as in others, custom has established the word, and it will remain as universally adopted.

The name is derived from ἄκρον, a summit or extremity, and μέγας (μεγάλη, μέγα), great. The name, therefore, designates a condition of hypertrophy or unnatural enlargement of the extremities of the body, viz., the hands, feet, fingers, nose, lower jaw and lips, ears and cranium, or, as Marie designated it, "une hypertrophie singuliere non congenitale des extremites superieures, inferieures et cephaliques." Von Recklinghausen proposes the term "pachyæmia."

ACROMEGALY

Symptomatology. — The first symptom which usually attracts attention is an unaccountable growth of the hands, and shortly afterwards of the feet. Women quickly notice the necessity for larger and larger sizes of gloves. Larger shoes are required; in some cases the sizes needed are extraordinary, reaching as high as No. 15 in the usual scale, the length of the foot attaining twelve inches.

The face undergoes corresponding changes. The nose broadens and lengthens, the lower lip and possibly the tongue become thickened and protrude; the head enlarges and the features become so coarse and unlike their former appearance that friends notice the marked change, or fail to recognize the individual after an absence.

The physician, however, is usually consulted for a persistent headache or for some visual defect, and in this way many of the cases have presented themselves at special clinics. In not a few cases they have been "picked off the street" by physicians who, thanks to Marie, have come to be familiar with the characteristic features of a disease that only a few years ago they allowed to pass unheeded.

The following classification of symptoms will be convenient in considering the subject:

I. Objective Symptoms.
II. Subjective Symptoms.
III. General Symptoms.

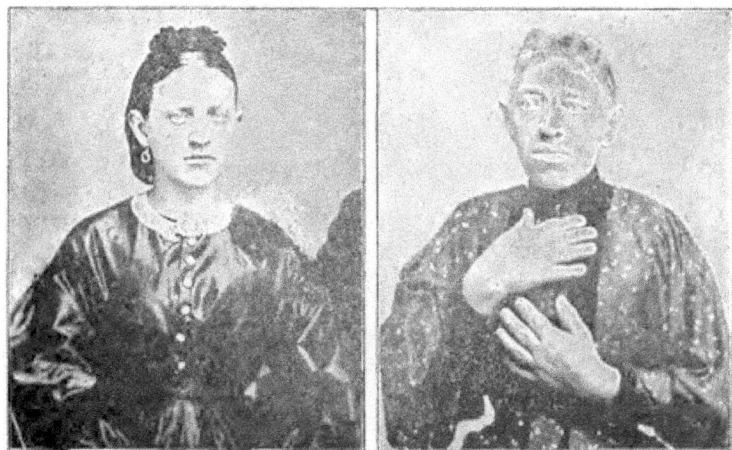

FIG. 2.

Dr. H. H. Vinke's case, aged 28 years.   Same patient, aged 51 years, fifteen years after onset of acromegaly.

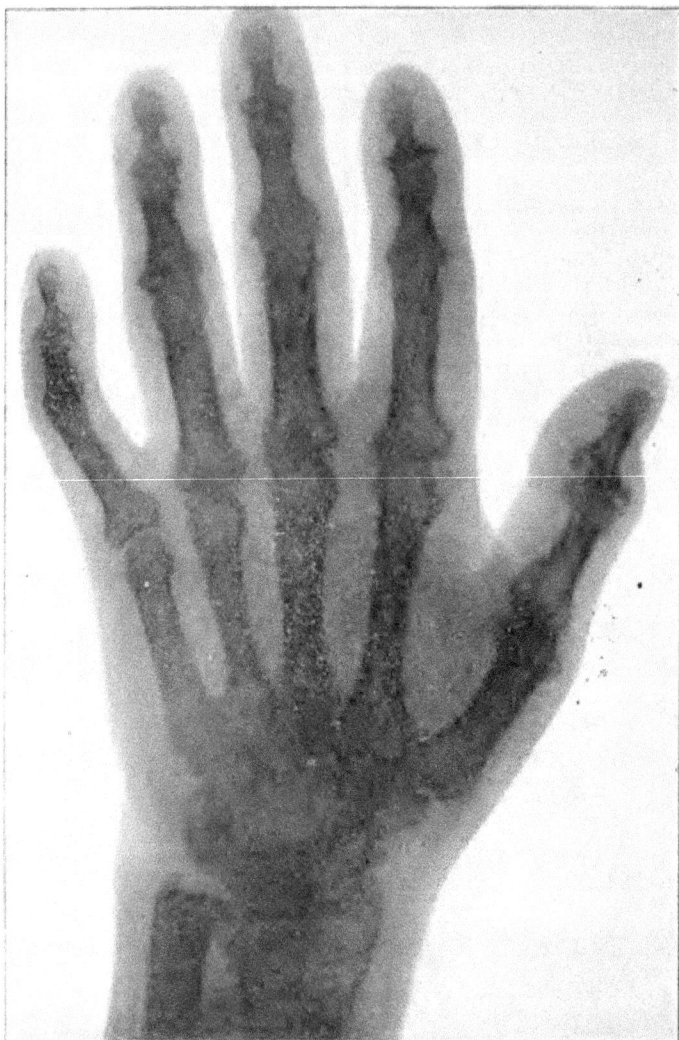

FIG. 3.—Skiagraph of right hand; case of acromegaly. Dr. Witmer's case; female, aged 55.

I. Objective Symptoms (fundamental): Hands; fingers; feet; toes; head; face; nose; lips; tongue; palate; chin; cranium; trunk; spine; thorax; ribs; sternum. Objective Symptoms (accessory): Neck; thyroid body; larynx; thymus; breasts; abdomen; genital organs; circulatory organs; respiratory organs; muscles; articulations; reflexes; electrical reactions; skin (sweat glands); urine; blood; sensory changes.

II. Subjective Symptoms (fundamental): Headache; amenorrhea; sexual desire; vision; thirst; appetite. Subjective Symptoms (accessory): Hearing; smell; taste; cardiac palpitation.

III. General Symptoms: Psychic depression.

OBJECTIVE SYMPTOMS.

Hands.—Among the first symptoms to be noted by the patient is an increase in the size of the hands. The customary gloves are discarded for larger and larger sizes, and in not a few cases it becomes necessary to have special gloves made. The fingers thicken and all the tissues seem hypertrophied. The fingers assume a shape commonly designated as "sausage-shaped." Ordinary rings cannot be worn. One of Virchow's patients had a special ring made that was so large that a German thaler could be dropped through it. In Roxburgh and Collis' patient, it is recorded that five times over she required to have her rings filed off her fingers and made of an increased size. On the palmar surface of the hands, and, for that matter, upon the feet as well, fleshy pads are liable to form. The epidermis and cutis become thickened, and the whole hand assumes a shape which is denominated "spade-like," or, as the French term it, "Main eu battoir" (battledore). This change in the hands is not so much due to a lengthening of the member as to an increase in circumference, which contrasts quite strongly with the unaltered arm and forearm. The nails are commonly flattened and enlarged and may be striated lengthwise. Nodosities have been noted about the phalanges.

Feet. — The feet share the same processes as are noted in the hands, and their increase in size is usually coincident with the hypertrophy in the upper extremity. Women note this with dismay, and the resulting deformity of the feet, as in the case of the hands, leads to much mental disturbance. The spirits droop, and anxiety on this account leads to melancholy and a desire for seclusion. The toes are thickened in the same manner as the fingers.

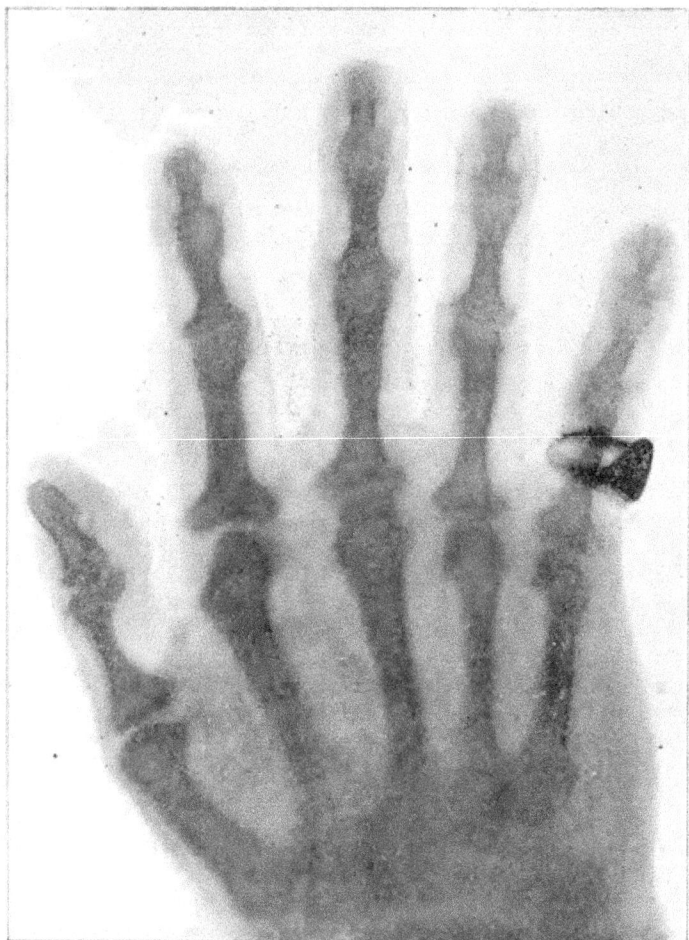

FIG. 4.—Dr. Packard's case of acromegaly described in vol. xvi of the *Transactions of the College of Physicians of Philadelphia*. This skiagraph should be compared with the picture of Arnold's case in *Ziegler's Pathology*. Hyperostosis of the phalanges is well shown. The entire outline of the ring on the little finger is plainly seen. Reduced.

The accompanying illustration shows the enormous size to which the feet may attain. The feet of this patient, who was 16 years old, measured 36 centimeters (14.25 inches); the width was 16 centimeters (6.25 inches); and the girth of the first toe 12.5 centimeters (5 inches).

In one of Dana's cases, a young giant, the feet were 35.5 centimeters (14 inches) long and the subject wore a shoe 46 centimeters (18 inches) in length.

A tendency to flat feet has been frequently noted, and thick cushions of soft tissues may surround the os calcis and outer border of the feet.

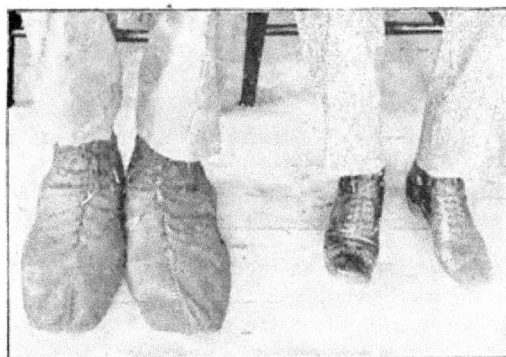

FIG. 5.—James W. Walker's case.

Skiagraphs of the enlarged hands and feet do not always reveal hypertrophy of bone. This is well shown in Sir William Broadbent's case, reported in 1896. The patient was a woman, and the bones are shown to be of normal size, the hypertrophy in her case residing wholly in the soft parts.

Head.—The head is commonly increased in size. This increase is in part due to an increase in the capacity of the sinuses, or the spaces between the two tables of the skull. This is quite evident in the skull of the case described by the author. The sinuses are capacious, although the tables of the bone are not elsewhere enlarged. In Regnault's case, however, the upper portion of the occipital bone was 14 millimeters in thickness.

The forehead is commonly low; the facial bones, with the exception of the lower maxillary, do not, as a rule, show any marked change. This bone, in a large proportion of cases, is the seat of characteristic changes. It is an "ἄκρον," or one of the extremities of the body. As the bone hypertrophies prognathism (cranium progeneum) ensues; the teeth in the lower jaw may project as much as three centimeters (1.25 inch) in advance of the upper teeth (Regnault's case). This is well seen in the accompanying Figs. 6 and 7, from Professor Osborne's case, in which a complete study of the skeleton was made. In the 130 cases of which we have made a synopsis, prognathism was mentioned as present in thirty-four. The changes in the lower jaw are commonly attended with a loss of teeth. Prognathism is the resultant of the growth of the body of the inferior maxilla, an increasing obtuseness of its angle and changes in its articular surfaces.

Nose.—This is one of the principal points of change. The nose becomes broadened and in many cases lengthened, and even a slight change in this regard makes great alterations in the general appearance. It frequently happens that the expression is so altered that it is difficult to recognize the individual after these changes have occurred. The nose flattens, the alæ thicken, and the nostrils are liable to be enlarged. The septum may undergo hypertrophy.

Lips.—The lips are commonly thickened; especially is this true of the lower lip, which may be greatly everted.

Tongue.—Occasionally the tongue is found to be enlarged, and this may be a source of considerable inconvenience. One of the

Fig. 6.

FIG. 7.

Normal skull.　　　　　Skull from case of acromegaly (Osborne)

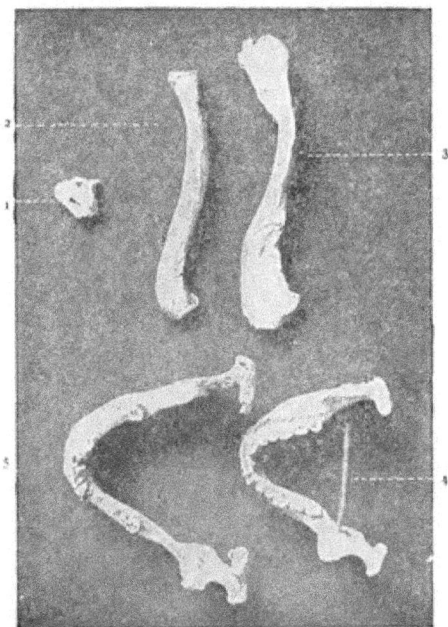

FIG. 8.—(1) Distal phalanx of toe; (2) normal clavicle; (3) clavicle from case of acromegaly; (4) normal inferior maxilla; (5) inferior maxilla from case of acromegaly. (Osborne.)

FIG. 9.   FIG. 10.

Jos. S——, aged 17. Showing: Eversion of the lips; prognathism (½ inch); lower jaw large. Irregular teeth; high palatal arch; narrow. Deformity of the head with frontal prominence. Total blindness, probably congenital. Exophthalmus; divergent strabismus; lateral nystagmus. Mentality good. Case shown to the author by Dr. F. Savary Pearce.

FIG. 11.—Enlargement of the nose in acromegaly (Curschmann).

early cases, that of Chalk, was described in 1857 under the title of "Partial Luxation of the Lower Jaw Due to an Enlarged Tongue." In our collection of 126 cases the tongue has been noted as enlarged in forty instances. The tongue, in extreme cases, projects from the

mouth and interferes with distinct articulation, and the natural tone
of the voice may be altered by changes in the tonsils, hard and soft
palate, and fauces. Even in women the voice may assume a deep
tone, and is variously described as "harsh," "guttural" (see case
of Chauffard).

Chin.—The lengthening of the chin is noticeable in the majority
of cases. This tends to give the face an oval contour. The prognathous chin contributes strongly to the change in the general
appearance.

Trunk.—The spine undergoes marked alterations as the disease
advances. One of the commonest symptoms is a cervico-dorsal
kyphosis. The head tends to fall forward upon the chest and, as in
Mossé's case, the bodily stature may be diminished. In the skeleton described by the author, it is estimated that the height would be
about three inches greater were it not for the spinal curvature
present. Kyphosis is due to changes in the bodies of the vertebral
and intervertebral substances, and it is not unusual to find ankylosis
or a bony deposit of a diffuse character, just as in the case of osteitis
deformans or osteo-arthritis. There is thus produced a bridging of
the superior and inferior margins of the vertebral bodies, due to the
ossification of the intervertebral ligament. Similar changes are
observed in rickets and acromegaly; they are liable to occur as
muscular weakness advances. In H. Alexis Thomson's case the
sternum and ribs were of gigantic size, but none of the measurements of his case approach the measurements seen in the case which
we describe.

The peculiar deformity due to kyphosis, taken in connection
with enormous hands and feet, may give an ape-like appearance to
the subject. Lavielle's case is a good example of this.

These changes in the thorax usually make the antero-posterior
diameter greater in proportion to the lateral. The sternum and
clavicles share in the hypertrophy. The abdomen, in well marked
cases, is prominent, sometimes pendulous even without obesity;
but the pelvis is not, as a rule, affected. Where the kyphosis is
extreme, the iliac crests may be in contact with the lower ribs.
This condition is sometimes described under the term "Polichinelle"
or "Polichinello," which was the name of an Italian humpback who
was famous in his day.

Long Bones.—The long bones are not, as a rule, affected, excepting in the giant form of acromegaly. An exception to this rule
was recorded by Leslie Thorne. The subject was a man of 48,
in whom the first symptoms had appeared at the age of 21. The

tibiæ and fibulæ showed a marked curve outwards and forwards. The long bones are more likely to show changes, if at all, in their extremities. This is shown very well in Osborne's case.

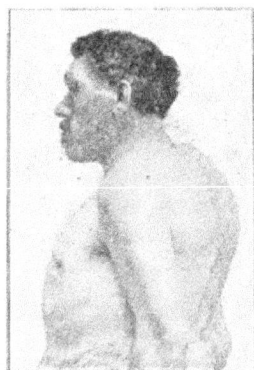

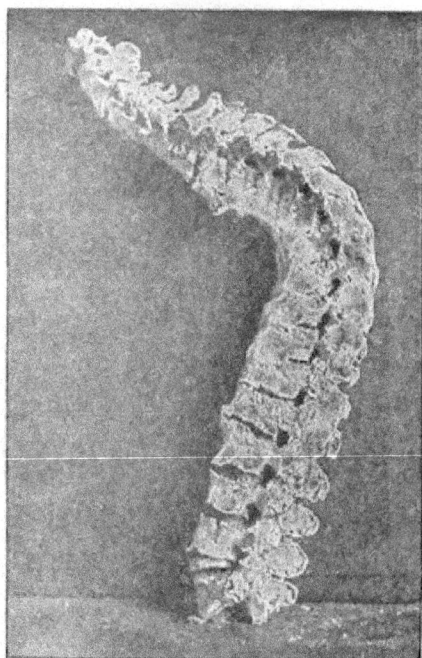

Fig. 12.—Deformity of thorax in acromegaly, and spinal column from same case (Osborne).

OBJECTIVE SYMPTOMS (ACCESSORY).

The neck is hypertrophied. We have found this noted twenty-eight times in our collection of cases. Tumors in the neck, entirely unconnected with the thyroid, have been observed by Henrot and Dreschfeld. In the latter case these tumors extended into the upper part of the thorax, where, to judge from the sternal dulness, a tumor existed, and it was believed took its origin from the thymus gland. The nature of these tumors was undetermined, but they were believed to be lympho-sarcomata, and it has been noted that growths of this nature sometimes take their starting-point from the remains of the thymus (Virchow-Koster). These growths naturally give rise to dyspnea.

Thyroid Gland.—This gland has been found to be either uni-laterally hypertrophied (Arnold), generally hypertrophied (Bailey, Godlee, Haskovec, Pechadre, Osborne, Verstraeten, Wolf, Carpenter, Furnivall, E. J. Smyth, Henrot), atrophied (Erb, Fratnich, Minkowski, Haskovec, Marie, Curschmann, Sigurini and Caporiacco, Tikhomiroff, Linsmayer, Somers), atrophied and sclerosed (Bonardi), absent on one side (Adler, Waldo), or normal (Freund, Hadden, Ballance, Comini, D'Esterre, Wadsworth, Goldsmith, Rolleston, Roxburgh and Collis, Strzeminski, Stroebe).

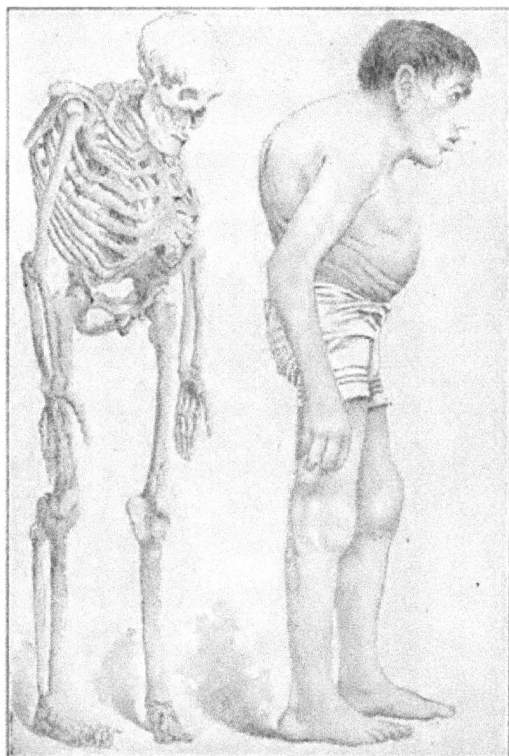

FIG. 13.—Skeleton of an acromegalic (Fritsche, Klebs, and Brigidi).

Although much attention has been paid to the condition of the thyroid in acromegaly, Dana remarks: "It does not seem to me

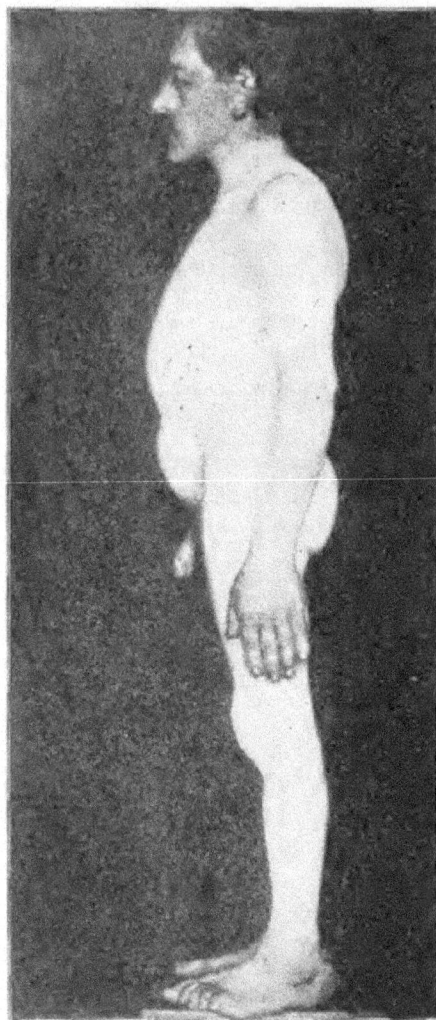

FIG. 14.—Pendulous abdomen in acromegaly (Curschmann).

that there is any evidence whatever that this gland has a relation to acromegaly." We cannot, however, dismiss the subject so summarily. The tendency is to ascribe considerable importance to the thyroid. Those who maintain that a relationship of importance exists are Erb, Gauthier, Haskovec, Chealde, Launois, Murray, Rogowitsch, Stieda, J. J. Putnam,[1] Wells.

Five cases of acromegaly are on record in which exophthalmic goitre coexisted. In another case an "extra-thyroid" gland was found by Osborne "in the median line high up in the chest cavity just above the upper end of the sternum." It contained "a large amount of iodine."

We think it very significant that in one case, at least, of myxedema the hypophysis has been found hypertrophied. This was in a case reported by Grón in 1894. It occurred in a woman of 62. The thyroid was atrophied, and at the autopsy the enlarged hypophysis occupied the entire sella turcica and was said to be "as large as a nut." It seems probable that this case was an example of compensatory hypertrophy. We believe that future investigations will doubtless establish a distinct relation between these glands.

Larynx.—The larynx shares in the hypertrophy of the neck. Even in a woman the cartilages become prominent in a few cases, and the tone of the voice grows deeper, or at least harsh, difficult, or naso-guttural. This change in voice is due partly to alterations in the cartilages and vocal cords, and partly to changes in the form and size of the antrum or other resonant cavities in the face. We have noted alterations in the larynx in eighteen cases. In a case recorded by Chappell the epiglottis was thickened, the arytenoid cartilages and the ventricular bands were enlarged, but the glottic aperture was very small. The pillars of the fauces, soft palate and uvula were thickened, and the tonsils and lingual glands were hypertrophied. While quiet respiration was free, but during excitement, labored. The patient died during an attack of dyspnea.

Thymus Gland.—The thymus gland has been supposed to play a prominent rôle in acromegaly. This opinion was first prominently enunciated by Erb, who laid great stress on hypertrophy of the gland, as revealed by dulness on percussion over the upper sternum. Schultze and Verstraeten have supported Erb in his belief as to the persistence of the thymus in acromegaly. "Erb's sign," as it is called, is rather uncommon in acromegaly. It has, however, been

---

[1] "Acromegalia probably stands as near to myxedema and diseases of the thyroid or kindred organs as Graves's disease, if, indeed, the bond be not closer."—Trans., Assoc'n Amer. Phys., 1893; paper on cases of myxedema, acromegalia, etc.

noted post-mortem by Arnold, Klebs, Dalton, Lathuraz, Mossé, Holsti, Rolleston, Ross, Roxburgh, and Collis, and clinically by Erb, Dreschfeld, Caton, Campbell, Bertrand, Flemming.

In Mossé's case, the post-mortem examination revealed an enormous growth or "*revivescence*" of the thymus. It extended in front of the trachea, from the thyroid gland to the pericardium, measuring twelve to thirteen centimeters vertically and eight centimeters transversely.

Mammæ.—The breasts are occasionally atrophied.

Genital Organs.—The external organs may be hypertrophied, while the uterus and testicles are found to be atrophied. It is a common thing to find that menstruation has ceased at the onset of the disease. In the giant form of acromegaly, as well as in the ordinary form, sterility is the ordinary condition.

Circulatory Organs.—These, as a rule, are only slightly affected, or, at least, little note has been made as to any changes in them. J. B. C. Fournier, however, has made a careful study of the heart in acromegaly, and has noted two forms of cardiac alteration: (1) A pure hypertrophy without degenerative lesion of the muscular fiber; this is generally the condition noted at autopsies. (2) A hypertrophy accompanied by cardiac sclerosis with degeneration of muscular fiber.

In the first instance there is a simple cardiomegaly accompanied exceptionally with insufficiency of the cardiac valves. In the second there is a true sclerotic myocarditis, a cardio-renal arteriosclerosis with what he terms a hypo-systole, a cardiac liver with edema of the feet, and albumen in the urine. Thus the heart participates in the general growth. In Osborne's case it weighed thirty-nine ounces.

Tachycardia has been observed by Ballard, Boltz, J. J. Putnam, du Cazal. Attacks of palpitation are occasionally noted.

The arteries share the fibroid change, becoming rigid, and the veins distend, especially in the lower extremities. Varicose veins are quite commonly observed.

The lesions of the vascular system present three phases—dilatation of the vessels, thickening of the walls, and obliteration of their lumen. So in the subcutaneous cellular tissue the vessels have their walls thickened, especially as regards the muscular coat; the inner coat is slightly thickened and not in all the vessels.

The Respiratory Organs.—These are not liable to be affected unless changes occur in the thyroid and thymus glands. The movements of respiration in aggravated cases are more abdominal

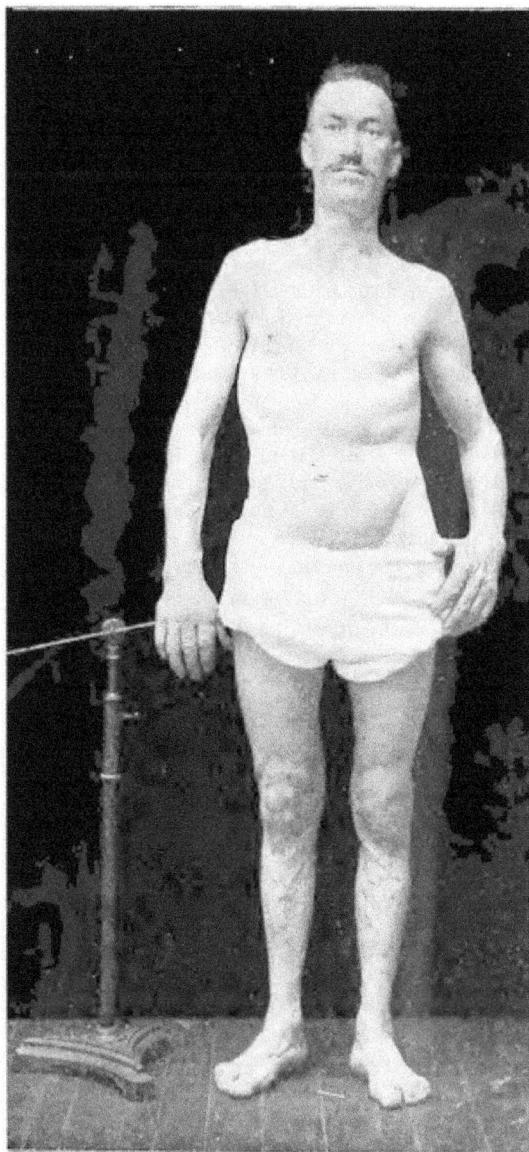

FIG. 15.—Varicose veins in acromegaly. Case shown to the author by Dr.

than thoracic; the upper thorax may be rigid and the spinal curvature may throw the thorax forward on the pelvis, as has been previously mentioned.

Muscles.—The muscular system varies in development with the type and stage of the disease. Naturally, the strength is great in the early period and particularly in the giant form. Virchow's patient is reported to have been able to lift 800 pounds; Eshner's case is that of a blacksmith in vigorous health, still following his arduous occupation. The fact that the disease has existed for a considerable time when the cases are reported, and that they apply to physicians for relief of symptoms, explains why muscular power is commonly noted weak as a rule. The muscles are not nearly so well developed as the measurements of the members would imply. The hypertrophy affects all the tissues, and muscular action is more likely to be impeded than facilitated. As a rule, the forearm remains unaffected, while the tissues of the hand increase in bulk. As the period of decadence sets in, muscular atrophy renders the patient quite powerless, and cardiac dilatation adds to the weakness of the circulation. Tremor is not unusual in acromegaly.

Joints.—The articulations are to some extent involved. This applies to the wrists, the smaller joints of the hands and feet. The changes in the intervertebral joints have already been referred to.

Reflexes.—These are generally preserved. In Stembo's and Osborne's cases the knee-jerk was absent on the right side and diminished on the left. It was absent in the cases of Remington, Pinel-Maisonneuve, Mendel, Claus, Banks, Marie, Pechadre, Flemming; nearly absent in Bertrand's case; weak and only obtained by reenforcement in the cases of Pick; weak in Massolongo's case, also in Hertel's, Dana's, Dercum's, Bignami's, Barclay and Symmes', Strzeminski's, Schultze's, Adler's, Haskovec's, Church's, and Hassert's; increased in the case of Pershing. In Marinesco's case it was normal in the left, but exaggerated and with clonus in the right. Thus in twenty-four cases, or 19 per cent., there was some change noted regarding the knee-jerks. The remaining 102 cases, as far as reported, were normal in this respect.

Electrical Reactions.—Conductivity to the electric current is increased, owing to diminution of the electric resistance. As the current is more readily conveyed by the blood-current than by other tissues, this would be the natural result. In this case, where the vessels themselves are enlarged, the resistance is much less than where the fluid is merely contained in the cellular tissue, as in edema. In the latter case there is no appreciable change in the

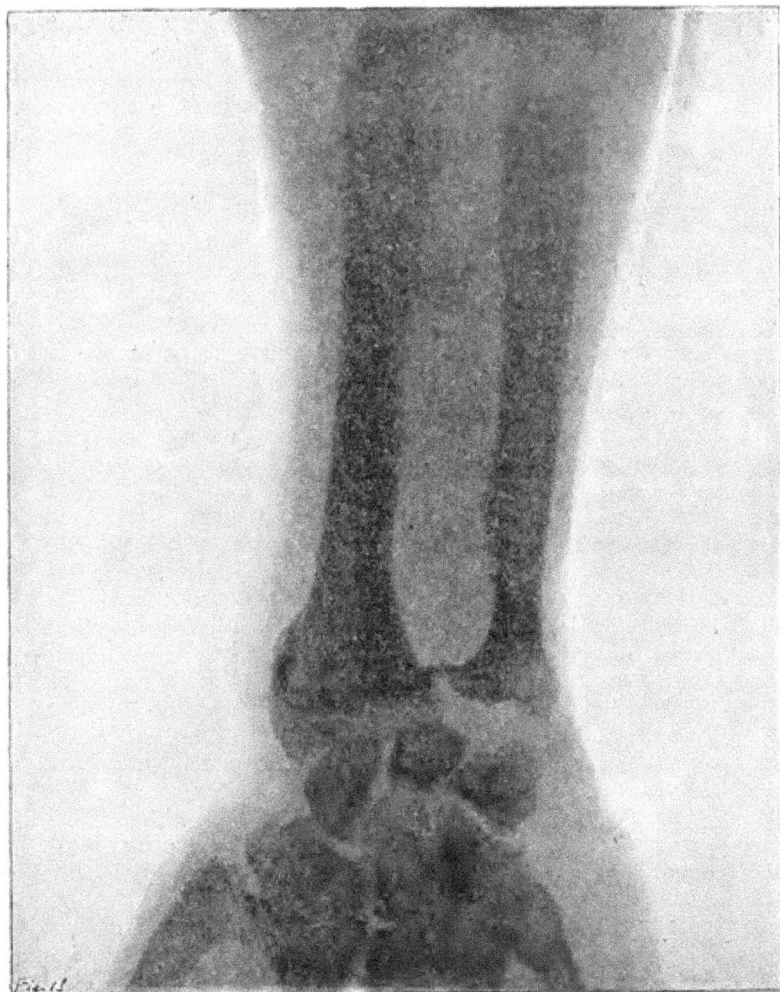

FIG. 16.—Skiagraph of right forearm and wrist from case of acromegaly (Osborne).

conductivity of the electric current. We have, then, every reason to expect in acromegaly a lessening of the electrical resistance as one of the symptoms. This has been denominated the "Charcot-Vigouroux" sign, and is present in exophthalmic goitre, in which, of course, the vessels are widely dilated.

Skin, Nails, and Hair. — These tissues undergo corresponding changes. In some cases the skin of the scalp is hypertrophied. The skin of the face is thickened and may be yellowish or sallow. Both the cutis and epidermis are thickened, the sudorific glands hypertrophy, and the pigment of the *rete malpighii* becomes excessive.

The general hue of the skin in advanced cases has given rise to the term employed by Gauthier, "cachexic acromegalique." It is not unusual to find reddish, pedunculated warts and mollusca fibrosa in acromegaly. The latter excrescences are quite commonly noted, and their presence has come to be one of the corroborative signs of the disease. The skin of the hands and feet may assume the character of coarse pads or folds. The hands are, therefore, of hard consistence, without pitting on pressure, as in edema.

The hair of the head may be coarse, and the nails may undergo striation either laterally or longitudinally; they may be short and flat and small in comparison with the hypertrophied fingers. The perspiration may be excessive and even offensive.

Acromegaly is thus entitled to a place among the dystrophies.

Urine. — The urine is commonly augmented in quantity. It is not unusual for a man having acromegaly to pass from 2000 to 3000 cubic centimeters daily. Pick's case passed 4000 cubic centimeters, Marinesco's 10,000 to 12,000 cubic centimeters, Walker's 12,600 cubic centimeters, in twenty-four hours.

Other cases where the urine was increased are those of Balzer, Banks, Barclay and Symmes, and Hanseman. This was also observed in the cases of Caton, Claus, Gordinier, Marinesco, and Moncorvo.

The urine often contains sugar. In the latter case the urine varies from 1036 to 1040 in specific gravity and contains 6.25 per cent. of sugar. In Ross's case the specific gravity was 1045. In Marinesco's case there was forty-eight grammes of sugar per liter.

Cases of glycosuria are recorded by Ballard, Brooks, Dalton, Dallemagne, Gajkiewicz, Gonzales-Cepeda, Marinesco, Ross, Strümpell, Squance, Walker, Thomson, Hanseman, and Lathuraz.

Albumen was noted in the cases of Ballard, Caton, S. Solis-Cohen, Arnold, Gauthier, Marinesco, Gajkiewicz, Dreschfeld.

Azoturia and phosphaturia may also be present.

Scanty urine has been noted by Bramwell twice, and by Hare. The urine was normal in 29 out of the 130 cases in our appendix.

Blood.—Changes have been observed in the few cases in which the condition of the blood has been examined and recorded. In Church and Hassert's case the hemoglobin reached 95 per cent. and the red corpuscles 7,000,000 per cubic millimeter. There were 400 red to one white corpuscle. In three cases of Marie and Marinesco the red corpuscles and the hemoglobin were both diminished. The blood of one of these patients was also examined by M. Lion in Professor Hayem's laboratory with similar result. Marie and Marinesco have also made the following statement: "Hyperplasia of the bone-marrow plays a part in this leucocytosis, because, at the same time, there were red corpuscles and eosinophile cells in larger quantity than normal."

Sensation.—Few changes in this regard have been noted. One of Marinesco's cases had normal tactile sensation in the left leg, but diminished in the right. In Chauffard's case there was a hysterical left hemianesthesia. Hypersensitiveness to cold was noted by Pershing and Gordinier. In Caton's case there was hyperesthesia over the left lower maxilla, and in Claus' case the tactile sense was noted as being acute. Subjective sensations are occasionally noted. Orsi's case complained of a sense of burning in the face. Haskovec's patient had regularly a sense of heat followed by great perspiration, succeeded by a sense of cold.

Temperature.—Verstraeten and Gauthier have noted a very slight difference between the axillary temperature and that in the palm of the hand. In Gauthier's case it was 36.5° in the axilla; 36.2° in the palm. The usual difference normally is 2° or 3° C. This is simply an indication of the excessive blood-supply of the extremities, a condition likely to give rise to hypertrophy of bone. Subnormal temperature was found in Waldo's case.

SUBJECTIVE SYMPTOMS (FUNDAMENTAL).

Headache.—This is one of the commonest complaints of the acromegalic. It is the direct product of the intracranial growth which is seated in the pituitary fossa. By its presence and its pressure on the optic nerves and the base of the brain this symptom becomes one of the cardinal signs of the disease. The anatomical relations of the hypophysis cerebri will be considered in another chapter.

Headache may be so persistent and severe as to affect the general disposition of the sufferer. The spirits are depressed and suicide has even been threatened, and carried out on account of the desperate condition of the patient (Pick). It may be constant or intermittent, and may be localized in the occiput, the frontal region, or confined to one-half of the head.

The duration of this symptom was, in Gauthier's case, twenty-nine years, and in Gonzales-Cepeda's case it was described as intense. In both of these the autopsy showed large growths of the hypophysis which amply explained the symptom. In Gajkiewicz's case the patient was described as groaning and lamenting for two years with headache. In Barclay and Symmes' case the duration was five years; in Ross's, three years, confined to the left temple. In one of Marinesco' cases it had lasted four years, and in another was seven in the occiput, while a third was confined to the temples. In Sanger Brown's case it was frontal; in Campbell's it was located in the eyeballs and head; in Caton's it seemed to be situated at the base of the skull.

Other cases where headache was a symptom are those of Banks, Barrs, Bertrand, Chauffard, Claus, two cases of S. Solis-Cohen (in one having lasted for seven or eight years), Dana, Dreschfeld, Eshner, Guinon, Goldsmith, Haskovec, Lavielle, Long, Massolongo, Mendell, Middleton, Olechnowicz, Orsi, Osborne, Packard, Parsons, Pechadre, Pershing, Phillips, Pick, Pinel, Ross, Rolleston, Schlesinger, Lynn Thomas, and Thomas of Geneva, Unverricht, and Whyte.

Intense neuralgia of the cranial nerves has called for surgical interference. Prominent among these is the case of Hare, which was operated upon by Keen (see under "Treatment"), and the case of Caton. Harris, Squance, Walker, Fratnich and Godlee have also recorded instances of severe neuralgia in acromegaly.

Amenorrhea.—This has already been referred to under the heading of "Genital Organs."

Sexual Desire.—*Pari passu* with the atrophy of the uterus and testicles, sexual power and desire abate, and as the disease advances are lost altogether.

Ocular Symptoms.—The appendages of the eye are liable to be involved early. The bones and orbital ridges are heavy; the cartilages and skin of the lids thicken by reason of a hypertrophy of the lowest layer of the true skin and of the connective tissue and gland of the skin. The secretion of tears is occasionally augmented. The pigment may be increased and the general color of the lids may be like bronze. The globes themselves are prominent, constituting an exophthalmus which has been so extreme as to occasion a true luxation of the eyeball (Hertel). In the cases of Motais and Pinel-

Maisonneuve, it is said that one could see the entire globe as far as the posterior pole.

We find exophthalmus recorded in 23 out of 130 cases, or 17.5 per cent. (Cases of Banks, Ballard, Benson, Bignami, Bullard, Brooks, Flemming, Gauthier, Guinon, Hare, Harris, Minkowski, Godlee, Hertel, Mossé, Motais, Orsi, Pinel-Maisonneuve, Strümpell, Thomas, Valat, Sigurini, and Caporiacco; see also case of Joseph S——, Figs. 9 and 10.)

Exophthalmus probably occurs by reason of a proliferation of the adipose tissue back of the bulb. One case was recorded by Burchard in which the eyeball itself was enlarged. The movements of the eyes may be slow, with a failure to raise the lids synchronously.

Nystagmus has been observed by Bignami, Boltz, Long, Maisonneuve, Packard, Whyte, Moncorvo, Schlesinger, and Strümpell. It may be either spastic or ataxic.

Internal strabismus has been observed by Banks, Benson, Campbell, Godlee, Moncorvo, and Lynn Thomas. Strabismus divergens has been recorded by Strümpell, Franke, Hertel, Litthauer, Marinesco, and Valat; of the right eye alone, by Bignami; and of the left eye, by Spillman and Haushalter, Schlesinger and Marinesco.

Thus there may be a complete oculo-motor paralysis (third nerve) and occasionally partial paralysis of the sixth. Diplopia may therefore be observed as a result of ocular paralysis. The irides sometimes react slowly to light, but normally to accommodation.

In Lynn Thomas' case the left iris did not react when a jet of light was thrown into the insensitive half of the left retina, it being a case of left temporal hemianopsia. But both pupils contracted when the light was focused on the sensitive portion of the retina (Wernicke's reaction).

Vision.—This has been recorded as normal in the cases of Appleyard, Barrs, Bertrand, S. Brown, Claus, Solis-Cohen, Dallemagne, Dana, Eshner, Fratnich, Gauthier, Gerhardt, Godlee, Gordinier (2), Guinon, Mossé, Osborne, Pick, Remington, Shiach, Squance, Thomas, Thomson and Waldo, Adler, Dalton.

Vision is impaired in the majority of cases. In our analysis of 130 cases, we have found some note as to defect of vision in about 61, or 48 per cent. It is quite likely that the proportion of cases of defective vision is much greater than these figures would imply, inasmuch as all cases are not studied with equal care in this respect.

The visual acuity may vary from normal to a slight amblyopia or even complete amaurosis, in which there may be either congestion of the optic nerves or neuro-retinitis with complete atrophy of the nerves. The form fields, and sometimes the color fields, are very commonly restricted. This condition may be an irregular narrowing of the visual fields or a clear-cut temporal or, rarely, a nasal hemianopsia (case of Bard).

Bitemporal hemianopsia is due to a blindness of the nasal half of each retina, causing a loss of the temporal half of each field. This symptom, as we shall see later, is caused by an overgrowth or neoplasm of the hypophysis pressing on the optic nerves and by bony changes in the optic foramina consequent on the abnormal growth of the lesser wing of the sphenoid bone. In this manner the arterial supply of the nerve and eyeball is disturbed.

Complete homonymous hemianopsia has been observed by Ross and also by Sir W. H. Broadbent, and was considered to be due to a tumor of the pituitary body pressing on the right optic tract. Left homonymous hemianopsia was seen by Dodgson, and was also recorded by Dulles, in the latter case the examination having been made by De Schweinitz.

An example of ocular changes in acromegaly is seen in a case recorded by M. Arthur Benson. The patient, a farmer of 38, first sought medical advice for failing vision. It was found that he had a central scotoma for colors and probably hemianopsia for color. These symptoms almost entirely subsided under the use of iodide of potassium and the cessation of tobacco. Two years later, however, when the use of tobacco was resumed, vision was totally lost; but gradually it was completely restored under the use of fresh thyroid extract, twenty-five minims three times a day. There remained, however, a slight defect in the upper temporal quadrant of the field in each eye—tetranopsia. The changes in this case were therefore largely of a toxic character.

More typical cases of ocular disturbance are those recorded by Strzeminski.

The visual disturbances are among the early signs and usually take a progressive course. In one of Schultze's cases the first symptoms of visual trouble had existed for ten years, and during the last half of this period there had been complete temporal hemianopsia. In Long's case blindness had existed for ten years, commencing at the age of 38. Complete blindness has occurred within two or three years from the onset of the ocular symptoms.

Bitemporal hemianopsia has been noted by Benson, Boltz, Bramwell, Sanger, Brown, Cowell, S. Solis-Cohen, Curschmann, Flem-

ming, Franke, Gajkiewicz (2), Hertel, Marinesco, D'Esterre,
Mendel, Packard, Schlesinger (2), Steinhaus, Strzeminski, Schultze,
Unverricht, and Valat.

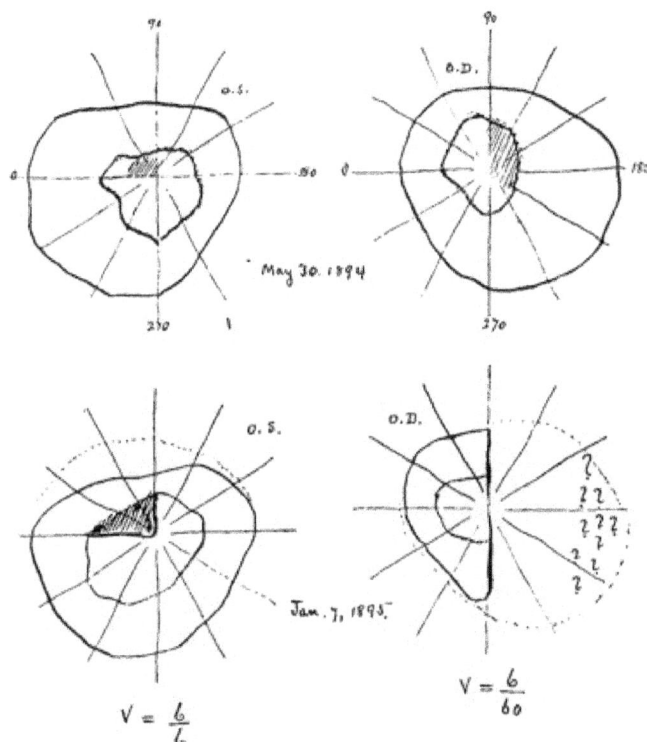

FIG. 17.—Fields from a case examined by Doyne.

Temporal hemianopsia of the right side is recorded by Banks,
Bramwell, Church and Hassert, also Mossé; of the left side by
Catou, Dercum, Hare, Pershing, Strümpell. Nasal hemianopsia
on the right by Roxburgh and Collis (see appendix).

The ophthalmoscope shows, in the early stages, a venous congestion; later on, a papillitis. In a case, as, for instance, that of
Benson, we find the nerve paler than normal with evidences of past
perivasculitis, while in Pel's case there was choked disk of such pro-
nounced type that the diagnosis of brain tumor was made from the

ocular examination. The condition is, however, a rare one, and its rarity is explained by Rask as due to a very marked agglutination of the optic nerve fibers under the pressure of the pituitary body, cutting off the communication of the intervaginal space of the nerve with the subarachnoid space of the brain, this communication being supposed necessary to produce choked disk.

Optic Atrophy.—This is, naturally, consecutive upon the pressure symptoms. Whenever, as we have learned from cases that have been examined post mortem, the optic nerves have been flattened by contact with the ever-expanding hypophysis, the microscopic examination will show a degeneration of the nerve fibers. The clinical symptom will therefore be, on ophthalmoscopic examination, a progressive atrophy of the optic nerve.

An observation recently made by Strzeminski (1897) shows us that in the cases of atrophy that are not preceded by temporal hemianopsia we must suppose that the gland considerably hypertrophied, and exerts from the start a pressure upon the nerves *in toto*.

Optic atrophy has been recorded by Bignami, Bramwell, Campbell, Caton, Chauffard, Cowell, Dallemagne, Denti, Dreschfeld, Costanzo, Flemming, Graham (2), Hadden and Ballance, Hare, Long, Massalongo, Packard, Rolleston, Roxburgh and Collis,

FIG. 18.   Pershing's case.   FIG. 19.

Schlesinger (2), Steinhaus, Surmont, Unverricht, Valat, Walker, Strzeminski.

How far the changes in the optic nerves are due to bony changes in the base of the skull has given rise to some discussion. It is probable, as Broca believes, that there may be some narrowing of the optic foramina from this cause; but evidence on this point is not at hand.

Pains in the eyeball of a neuralgic type are recorded by Erb, Lombroso, Hare, Tanzi, Strümpell.

Thirst.—Patients sometimes have an inordinate thirst, a demand for fluid out of all proportion to even the extraordinary size of some of the subjects. Remington's patient could drink a gallon of beer at once; but he was a large man, six feet two inches in height, and weighed 268 pounds. One of the cases which we record could drink three or four quarts of milk at once. We find a memorandum of 17 patients among 130 in whom polydipsia was a symptom. Increased thirst is doubtless connected with the glycosuria present in many of these cases.

Appetite.—This is increased more frequently than thirst and is recorded in twenty-five cases. This symptom may exist without increased thirst, and *vice versa*. In one case it was noted that eating gave great relief to the general discomfort of the patient. Bulimia is thus one of the characteristic signs of acromegaly.

SUBJECTIVE SYMPTOMS (ACCESSORY).

Hearing, Smell, and Taste.—The special senses are not generally disturbed, but in exceptional cases affections of smell and hearing are noted. Hearing may be disturbed by bone growth and tinnitus aurium may be a very troublesome symptom. (Cases of Osborne, Barclay-Symmes, and Holsti.)

Cardiac Palpitation.—This has been referred to under the heading of the organs of circulation. It is present occasionally, especially in cases where enlargement of the thyroid or thymus is found, or where the great increase of bodily weight, in conjunction with muscular fatigue, overtaxes the heart.

GENERAL PSYCHIC SYMPTOMS.

Depression.—As mentioned at the beginning, the extraordinary changes that occur in the physiognomy and in the hands and feet are a source of great mortification and mental distress to the unfortunate subjects. The mind is naturally affected, and depression of the spirits is the natural outcome of the progress of this grotesque

disease. The temper may be irritable, and impulses of a suicidal or
homicidal nature are recorded (Osborne, Pick, Haskovec). The
memory is, as a rule, unaffected. In cases in which the headaches
are violent and of long duration, and in which muscular feebleness,
lassitude and inability to pursue the usual avocations mark the
advance of the disease, it is but natural that profound mental
depression should ensue. In the latest stage coma ushers in the
fatal termination.

The Nervous System.—In a few cases we have organic changes
which may be more or less connected with the disturbances of
the pituitary gland. Epilepsy is one of these and has been
observed by Marinesco. Drs. Raymond and Souques have reported
a case of acromegaly of many years' standing in a man aged 54,
who in the last three years had developed Jacksonian epilepsy lim-
ited to the right upper extremity and right side of the face. They
stated that the pituitary gland, in their opinion, constituted a cerebral tumor capable of exciting from a distance the cortical psycho-
motor centers.

Marinesco's case was that of a woman aged 32, who had had
epilepsy at 22, the attacks occurring three or four times a week.
At the age of 25 she had an attack of giddiness, in which she fell
from a second floor. Six months later she noticed that her feet
were enlarged, and later her hands, face, and abdomen. The
menses ceased. Strabismus, polyuria and weakness supervened.
Sensibility was preserved in all forms except vision. Examination
revealed bitemporal hemianopsia. The urine contained forty-eight
grammes of sugar per liter.

Paraplegia has been observed in acromegaly and was due to
pressue of the tumor on both motor tracts in the crura cerebri. In
Pershing's case there was also a loss of bladder and rectal control.
Syringomyelia has been found coincident with acromegaly by
Holschewnikoff, Frederick Peterson, and Bassi. In the first case,
however, it should be noted that there was no affection of the thyroid and thymus or of the pituitary body; although, on the other
hand, the hands were large and hypertrophied and the skin covering
them was thickened in the superficial and deeper layers, and the
bones of the hands showed small exostoses. The second case has
not yet come to autopsy.

Bassi's was a remarkable case, in which syringomyelia and
solitary tubercle in the cerebellum were associated with cephalic
acromegaly. It is greatly to M. Bassi's credit that he was able to
diagnosticate all three of these affections *intra vitam*, and he was

fully corroborated at the autopsy. It would appear that the syringomyelia antedated the acromegaly and cerebellar tubercle.

Glioma of the brain and apoplexy were noted by Pel in a case of acromegaly.

Tinnitus may be caused by the pressure of the growth upon the carotid arteries.

Paresthesias and pains in the extremities, according to Maximilian Sternberg, are frequent in acromegaly. On this account many cases have been treated and considered as cases of rheumatism. Pressure on the brachial plexus and the trigeminus due to bone changes has been noted (Ascher).

Great sensitiveness to cold is recorded in several cases, and, on the other hand, one case, that of Rolleston, had a subjective sense of heat even in the coldest weather and felt obliged to wear thin clothing. The pains are consequent on the extraordinary tissue changes and are sometimes described as giving the sensation of the bones being torn apart.

The symptom-complex which we designate as acromegaly may be tabulated in the following manner, as M. Sternberg has done, or in the synoptical chart which has been used in our analysis of the cases comprised in the appendix (not included in the present publication):

1. Objective symptoms: (*a*) *Constant.*—Enlarged hands and feet; lengthening of face; enlarged eyelids; excessive enlargement of nose; prominence of cheek bones; enlarged lips; enlarged chin; prominent jaw; kyphosis; thickening of bones of thorax; abdominal respiration. (*b*) *Inconstant.*—Prominence of supraorbital arch; exophthalmus; optic nerve atrophy; hemianopsia; impaired hearing; anosmia; disordered taste; enlarged larynx; depth and roughness of voice; Erb's dulness; atrophy of testicles; enlargement of penis; enlargement of abdomen; atrophy of muscles; reflexes; enlargement of heart; increased rate of pulse; varicose veins; enlarged lymphatic glands; impotence; sweats; polyuria; glycosuria; disordered sensibility; pigmentation; warts and moles.

2. Subjective signs: (*a*) *Constant.*—Loss of sexual instinct; polyphagia; polydipsia. (*b*) *Inconstant.*—Headaches; palpitation; dyspnea; paresthesia; vasomotor neuroses.

3. General and psychic symptoms: General weakness and ennui; melancholia.

Varieties of Acromegaly.—Three distinct varieties or forms of acromegaly have been differentiated by M. Sternberg. They are as follows:

1. A benign form. Duration, about fifty years; changes, slight.

2. The usual chronic form. Duration, eight to thirty years.

3. An acute malignant form. Duration, three to four years. In this form the pituitary is sarcomatous.

COURSE—DURATION.

The course of acromegaly is essentially slow. In its onset it is insidious, and for that reason it is always difficult or impossible to fix the exact date of its inception. In the case of women we may be guided somewhat by the time at which menstruation ceased, and in this sex any change that may occur in the size of the hands and feet will be more readily noted and remembered. There is difficulty, also, in making any positive statements as to the arrest of the disease, should this occur. It is found that five or ten years or even longer may elapse before menstruation is disturbed.

There is some variation as to the member which is first affected. Although the hands are probably the first to be affected, in one case (Kanthack's) the starting-point was the left foot, and the second toe grew to an enormous size. In Tamburini's case the disease started in the feet, involving next the head and then the upper extremities. The occurrence of prognathism, while common in acromegaly, is not always noticed by the patients.

Figs. 18 and 19 show the general progress of the disease. The first photograph was taken eight years before the disease first appeared; the second was taken about fifteen years after its origin, but, as we have intimated, its onset was so gradual and insidious that neither she nor her relatives can state accurately just when the trouble began.

Although the progressive nature of the disease must be recognized, it is quite likely that arrest will be more frequently recorded as its nature and treatment are better understood. The case of Costanzo and a recently published case by Vinke are probably as encouraging as any that have been presented. The latter patient was treated with thyroid and pituitary extracts, alone and combined with each other, and remained under treatment for five months. The results in both instances are recorded in the chapter on Treatment.

It would thus appear that appropriate treatment may be followed by an actual diminution in the size of the hypertrophied limbs.

Two stages in the course of acromegaly have been differentiated, viz., the erethic stage and the cachectic stage (Gauthier).

The phenomena of erethism which characterizes the first stage embraces, first, a painful hyperesthesia which manifests itself in headaches and rheumatic pains; second, a hypertrophy of the muscular fibers which may give to patients a muscular power greater than usual; third, palpitation of the heart accompanying the hypertrophy of that organ; and finally the polyphagia and polyuria which may be considered to be connected with an erethic state of the respective organs.

The second stage is characterized by a cachexia or a period of decadence. The stage of increase has abated and the phenomena of erethism have disappeared. Muscular atrophy and cardiac dilatation and a consequent enfeeblement of the circulation render the patient quite helpless. It is in this stage that bleeding from the nose may ensue, and progressive debility marks the period of decline, which ends in syncope. Epistaxis may also occur early.

Tamburini has designated the stage in which the visual functions are interfered with and in which headaches develop as "*la phase hypophysaire.*"

The duration of acromegaly may cover as much as ten, twenty or thirty years. When death occurs it is usually by syncope terminating a progressive cachexia. Death has occurred, however, during an epileptic convulsion.

ANATOMY.

The discovery of acromegaly has awakened great interest in the anatomy and physiology of the pituitary body. Indeed, it was not until 1890 that the announcement was made that the pituitary body is the seat of such marked alteration as to suggest that it is the *fons et origo mali.* To Souza-Leite belongs the credit of this observation. Up to that point, the pituitary body, or *hypophysis cerebri*, as it is termed, was regarded with indifference. Descartes had made its neighbor, the pineal gland, world-famous, as the supposed anatomical seat of the soul; but the pituitary body, as that name implies, was relegated, in the minds of anatomists, to the humble duty of secreting the nasal mucus. It remained contented with this ignominious office until its emancipation in the early part of the present decade, and during these latter years it has attracted universal attention and assumed gigantic functions.

The pituitary body, *hypophysis cerebri* (German, Hirnanhang), is a reddish-gray, extremely vascular mass, situated in the sella turcica. It is oval in shape and weighs from five to ten grains.

Boyce has made a careful study of the gland, based on an examination of one hundred brains of insane subjects, dying of intercurrent affections. The weight averages .5 gramme, ranging usually from .3 to .6 gramme, the size, apparently, bearing no relation to age, and more particularly to sex or general nutrition, or to the size of the brain.

The hypophysis is held in place by a process of the dura derived from the inner wall of the cavernous sinus. There is a small hole in the center for the passage of the infundibulum.[1]

The two lobes are separated from each other by a fibrous lamina. The anterior is somewhat the larger, oblong, with a concavity behind, in which the round posterior lobe is received.

The anterior lobe, of a dark yellowish-gray color, is developed from the ectoderm of the buccal, or primary oral, cavity, and resembles somewhat, in microscopic structure, the thyroid body. It consists of isolated vesicles and slightly convoluted tubular alveoli lined by columnar or ciliated epithelium and united by loose vascular connective tissue.

A majority of the acini are solid; they occasionally contain a colloid material, similar to that found in the tubules of the thyroid body. Their walls are surrounded by a close network of lymphatics and capillary blood-vessels. Only a few of the acini at the edge of the smaller lobe are hollow. This portion of the gland is related to the alimentary tract, and it is the part that is especially liable to disease.

The posterior and smaller lobe is developed by an outgrowth from the embryonic brain and during fetal life contains a cavity which communicates, through the infundibulum, with the cavity of the third ventricle. In the adult it becomes firmer and more solid, and consists of sponge-like connective tissue arranged in the form of reticulating bundles between which are branched cells, closely resembling bipolar or multipolar ganglion cells. This portion is evidently vestigial or rudimentary in man.

In the lower animals the two lobes are quite distinct, and it is only in the mammalia that they become connected. The hypophysis thus assumes a much more prominent feature in the brains of the lower animals than in the case of man. It is not strictly a ductless gland, but its secretion is poured through an imperfect system of ducts opening between the dura and pia mater.

The accompanying illustrations show the position and comparative size of the embryonic hypophysis in the sheep and the chicken.

[1] For the anatomical description, see Gray's Anatomy and a description by Philip Stohr.

The pituitary and thyroid bodies start as pouches from the wall of the oral cavity—but growing dorsad, the hypophysis is embraced between the ossifying centers at the base of the skull and is thus included within the cranial cavity. Andriezen states that it originally " poured its secretion into the current of a water-vascular system which, beginning at the mouth, irrigated the cerebral ventricles and central spinal canal of the earliest vertebrates. The main function of this system must have been respiratory, but the pituitary gland probably furnished some substance necessary for the proper nutrition of the nerve tissue, which it now nourishes less directly through the blood and lymph " (Pershing).

Andriezen, after tracing the development of the pituitary body in the lower vertebrates through the amphibia to mammals and man, shows that its related size does not keep pace with the increased growth of the nervous system, an indication probably that having attained the acme of its activity in lower vertebrates, it is, in higher forms, already beginning to show signs of diminishing activity, though, of course, still functionally active even in man.[1]

There is, as Andriezen states, an obvious parallelism between the thyroid and the pituitary, not only in their early evolution in vertebrate animals, but also a physiological relationship. Thus we can explain the enlargement of the pituitary which has been observed by Stieda, Rogowitsch, Hofmeister, and Gley, after thyroidectomy. But Andriezen notes that it must be remembered, however, that " the pituitary belongs both anatomically and physiologically to the central nervous system, while the thyroid belongs to the respiratory function of the blood-vascular system and thereby to the tissues generally."

All of these investigations point to the important trophic influence of the pituitary gland on the central nervous system. Just how this operates, we are as yet unable to state; even the observations on the experimental destruction of the pituitary by Horsley in 1886, and by Gley, Marinesco and others in 1892-93, appear negative; but such experiments are liable to defects which do not permit too positive conclusions. Granted that Andriezen is correct in his own conclusions that the pituitary gland exercises a trophic action

[1] Rogowitsch has found that after thyroidectomy in dogs, the pituitary gland becomes swollen, the cells vacuolated and ultimately disintegrated. Bourneville and Brecon state that in sporadic cretinism, where the thyroid gland is lost, the pituitary body has been found to be enlarged.

Boyce and Beadles found, in a case of myxedema, that the acini and extra-acinar tissues were filled with colloid-like material cells and that the cells were increased in size. Abram found changes in the pituitary in a case of carcinoma of the thyroid.

× 10

Lat. ven.
chor. plex
Striat
3rd. Ven.
Med. oblong
Plexus
of roof

Sheep.

× 3½

Chick

× 10

Hemisphere Brain
chorid. Plex
thalamic Brain
Mid. Brain
Striatum
cereb.
Craniel Flexure
3rd. Ventricle
Nose
Pons
Sup. max
Tongue
Medulla
Thoracic Wall
Inf. max
Pituit Body
cochlea
ganglia

Sheep

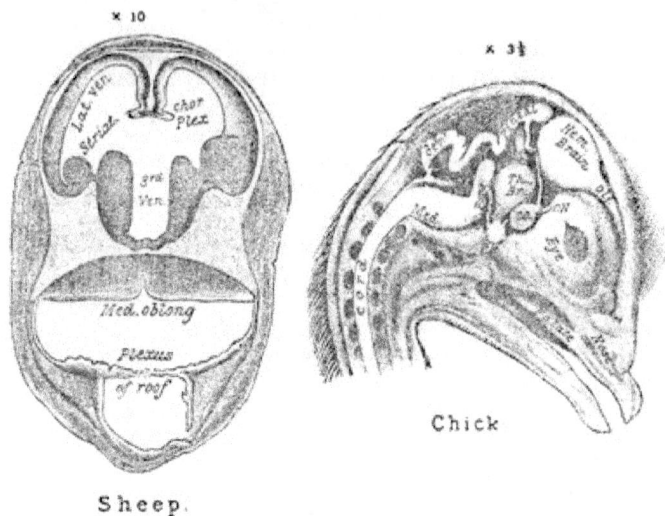

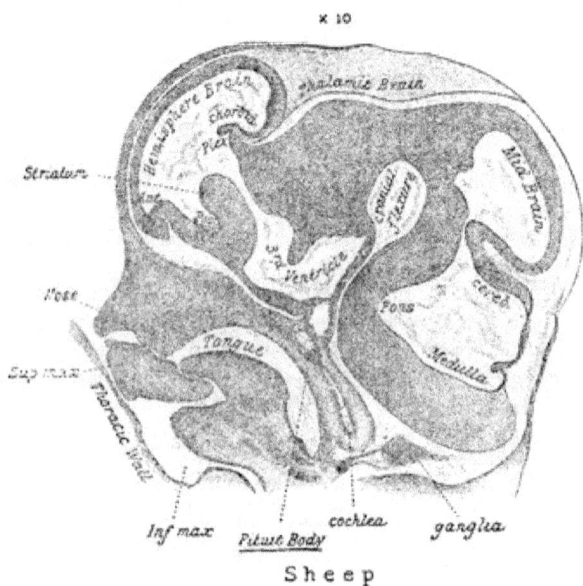

FIG. 20.—(After Dr. Alec Fraser.)

on the nerve tissues, enabling them to take up and assimilate oxygen from the blood-stream and to destroy and render innocuous the waste products of the metabolism, and these two functions are intimately related—in fact, really one process—so that an adequate assimilation of oxygen by the nerve tissues secures an adequate destruction (by oxidation) of the waste products.

Thus we may expect, *a priori*, that the results which would follow removal or destruction of the pituitary body would be those due first to a malassimilation of oxygen by the nerve tissues, and simultaneously, in the second place, to an insufficient destruction and consequent accumulation of waste products. In this way there ensues a rapid nutritional failure and death of the central nervous system. The results would, therefore, be:

1. Depression and apathy (the commencing failure of activity in the nerve centers).

2. Muscular weakness, the first peripheral effect.

3. Loss of fine coordination and equilibrium.

4. The development of twitchings and irregular contractions or spasms of the muscles.

5. Deficient heat production and subnormal temperature.

6. A wasting of the body tissues, in relation to the more rapid failure of the central nervous system.

7. A probable compensatory polypnea or dyspnea, the peripheral indication of the failure of the nerve centers to assimilate oxygen.

8. A rapid progress toward death.[1]

PATHOLOGY.

Although, as we have noted, the hypophysis has been observed slightly enlarged in myxedema and cretinism, it remains true that in acromegaly we have the only disease in which the pituitary attains any considerable size. Gauthier remarks that the hypophysis is a veritable ἄκρον and, consequently, especially liable to hypertrophy, just as in the case of other terminal points in the economy. I, however, do not lay much stress upon such a construction of the anatomy of the hypophysis, as I believe whatever pathological changes it undergoes are entirely referable to cellular and developmental characteristics and have no relation to gross physical form.

We have to deal with the remarkable fact in acromegaly that the pituitary body has a tendency to increase *pari passu* with the

---

[1] For a fuller account of the histology of the pituitary body, reference is made to the works of Luschka, Langen, Flesch, Dostoiewsky, Lothringer, Rogowitsch, Andriezen (1894), Berkley (1894), Schonemann, Stieda, W. Krause (1876), W. Müller (1871), and Haller (1896).

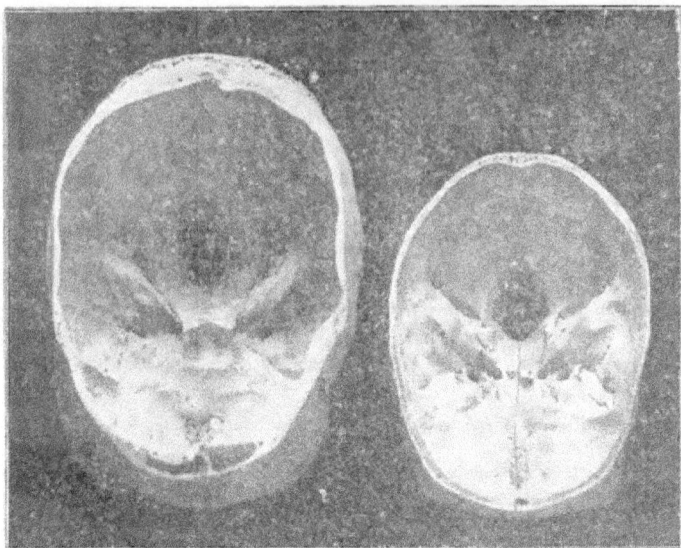

FIG. 21.—Base of the skull of the American giant, showing enlarged pituitary fossa, compared with normal skull from the collection of Dr. George McClellan.

development of the affection. The expansile power, coincident with hypertrophy, is able to deform and hollow out the bone by reason of its great vascularity and proliferative power. The sella turcica is dilated into a cavity of corresponding size. In an extreme case the growth presses on the optic commissure, the cavernous sinus, and on the carotid artery; on the third and fourth nerves; slightly on the ophthalmic division of the fifth nerve; on the uncinate convolution of the temporo-sphenoidal lobe; in some cases, on the olfactory tract; on the posterior surface of the orbital convolutions; on the under surface of the crura; on a part of the internal capsule and thalamus, and even on the anterior surface of the pons.

In Roxburgh's and Collis' case the enlarged pituitary had so pressed upon both optic tracts and the chiasm as to cause a total disappearance of the left tract and partial destruction of the right. The dura of the sella turcica presented an irregularly eroded and vascular appearance, and the irregularities were filled with an extension of the pituitary growth.

Enlargement of the pituitary fossa is, in general, a fair means of determining the size of the pituitary body. In my case (the giant) the extreme antero-posterior diameter is 2.7 centimeters as against .8 to 1.2 centimeters in the normal skull. This is 4.5 millimeters larger than Thomson's and Broca's cases, and the lateral diameter is 42 millimeters as against 28 millimeters in the normal. In Rothmell's case the pituitary body was enlarged to the size of a hen's egg and weighed 476 grains (32 grammes).

We cannot overlook the fact, however, that dilatation of the pituitary fossa by itself is not a certain proof of acromegaly.

There is, in the Musée Dupuytren, a skull in which this fossa measures eighteen millimeters in antero-posterior diameter in which there was evidently no acromegaly, as the face is normal and there is no hypertrophy of the bones of the skull. The specimen originally showed a tumor lying between the brain and the skull.

Sarcoma and Other Tumors of the Pituitary Gland.—In one of Mossé's cases the hypophysis was found to be replaced by a "fasciculated sarcoma" weighing 36 grammes (540 grains). The thyroid gland had undergone cystic degeneration; the thymus, however, had hypertrophied and assumed a pyramidal form.

In Strümpell's case, reported in June, 1894, and which had been studied for a long time in the clinic at Erlangen, a tumor of the hypophysis was found having the character of a malignant sarcoma.

In Caton's case there was a tumor the size of a Tangierine orange, occupying a much dilated sella turcica and, apparently, sarcomatous.

Adenoma of the hypophysis was found in the cases of Pearce Bailey, Boltz, Stroebe, and Tamburini.

Boyce and Beadles, Dallemagne, Gauthier, Marino, and Osborne found cystic degeneration. In the description of Roxburgh's and Collis' case it is stated that, while most of the tumor of the hypophysis had the appearance of a simple hypertrophy, the peripheral and especially the basal portions presented characters midway between those of a glioma and a round-cell sarcoma.

In autopsies in acromegaly the hypophysis has always been found to be enlarged or diseased. This position has been challenged on insufficient ground, but is now coming to be accepted as correct.[1] My own independent study and conclusion in this respect is emphatically corroborated by Maximilian Sternberg, who has recently made the unqualified statement: "Die hypophysis . . . ist bei Akromegalie stets erkrankt." He cities forty-seven autop-

[1] Average weight of hypophysis of man .5 to .6 gramme.

sies in proof. The list which we present numbers fifty-seven autopsies, the largest number hitherto placed on record.[1] They are as follows:

Bailey, 2.2 x 1.6.
De Berg.
Boltz (1).
    (2).
Bonardi.
Bourneville and Regnault (Case I).
Boyce and Beadles.
Brigidi, 2.8 x 3.8 cm.
Broca.
Brooks' and Baruch's, 1.5 x 7 cm.
Bury.
Caton and Paul, size of Tangierine orange.
Cepeda.
Claus.
Comini.
Dallemagne (1), size of pigeon's egg.
    (2).
    (3), size of nut.
Dalton.
Dana, 3 x 3 cm., weight 2.
Duchesneau.
Fraentzel.
Fratnich, 5 x .
Fritsche and Klebs.
Furnivall.
Gauthier.
Godlee, size of cherry.
Griffith.
Hanseman.

Henrot, 3 x 4 2 cm.
Holsti, 2.5 x 3 cm.
Lancereaux.
Langer-Sternberg.
Linsmayer.
Marie and Marinesco.
Mossé and Daunic, weight 3.6.
Osborne.
Packard, 2.5 x 3 cm.
Peehadre and Lathuraz.
Pineles.
Rolleston, size of walnut.
Rothmell, weight 32, size hen's egg.
Roxburgh and Collis, size of walnut.
Schultze, 4 x 2 cm.
Sigurini and Caporiacco, 11 x 7 x 6.
Smyth, E. J., 5 x 7.5.
Squance.
Stroebe, 2.5 x , size of nut.
Strümpell, 2 x 2.5 cm.
Tamburini, 5.3 x 3.9, weight 2.
Tikhomiroff.
Thomson, size of walnut.
Uhthoff.
Verga.
Wolf, K. (1).
    (2).
Worcester, weight 58, length 4.6 cm.

DIMENSIONS OF PITUITARY FOSSA IN SKELETONS.

Regnault,	3.1 x 3.6 ....
American Giant (Author's case).	2.7 x 4.2 x 1.2
Irish Giant	3.8 x x 2.8
Byrne, R. C. S. E.	2.2 ....
Hutchinson, W.	3.1 x 3.8 ....

In Duchesneau's case the hypophysis contained a wine-colored fluid, which was attributed in part to the fact that thirty-six hours had elapsed at the time of the autopsy. Claus shows that in his case the autopsy was performed between two and three hours after death, and the fluid found could not be accounted for by cadaveric liquefaction, but must have been due to a process of degeneration of the affected gland.

In about 12 per cent. of specimens of hypophysis in acromegaly the disease is a true sarcoma.

Tumors of the Hypophysis Without Acromegaly. — Drs. J. M.

[1] We intentionally omit the cases of the Hagner brothers and that described by Sarbo; also the case of Somers. In the latter case no examination was made of the brain, and therefore it is not properly included in lists of autopsies in acromegaly.

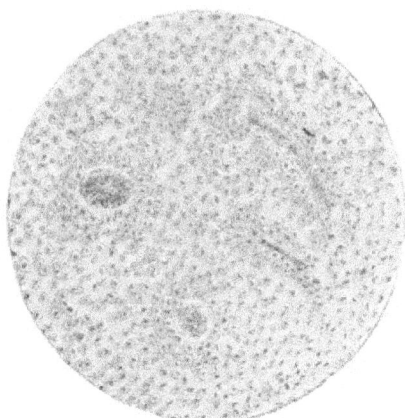

FIG. 22.—Sarcoma of pituitary in a case of acromegaly (Strümpell).

Anders and Henry W. Cattell, in 1891, reported to the Philadelphia Neurological Society a hemorrhagic tumor of the pituitary body and infundibulum, unattended with any signs of acromegaly. The symptoms in life had been anemia, dizziness, facial neuralgia, internal strabismus, and general weakness and nephritis. Death followed convulsions.

In a case which occurred in the practise of Weir Mitchell a large aneurism of an anomalous artery crossing the sella turcica entirely obliterated the hypophysis, but no symptoms of acromegaly ensued. It might be urged, however, that in cases of this class a deficiency of the pituitary for a longer period of time might be followed by symptoms which, it is well known, require considerable periods for their development.

Reports by Weir Mitchell and other observers bear out the statement that we may have tumors of the hypophysis without signs of acromegaly. Pearce Bailey has recently reported a case of parenchymatous hypertrophy of the pituitary with death due to hemorrhage into the gland structure. There had been headache, dimness of vision, blindness, and paralysis, but not acromegaly.

I have recently observed a fatal case of pleuro-pneumonia in which the autopsy revealed a large tumor of the hypophysis, apparently sarcomatous; the case presented no symptom whatever of acromegaly, either clinically or pathologically. It occurred in a colored man, a native of North Carolina, 38 years of age. After a short illness he died of double pleural effusion, hypostatic pneumonia, and pulmonary edema. The heart was hypertrophied; there was pericardial effusion, together with interstitial nephritis, arteriosclerosis, and fatty infiltration of the liver. The pia mater showed no lesions; dura and calvarium normal. There was no lesion of the convexity of the brain. Occupying the position of the pituitary body was a tumor measuring 3.5 centimeters laterally and three centimeters antero-posteriorly. It had the firmness of brain tissue and fitted the enlarged sella turcica. The walls of the sella were not eroded, excepting as regards the anterior clinoid process on the left side. Microscopic section showed the growth to be a round-cell sarcoma.

In the experiments which have been undertaken to see whether acromegaly will follow the destruction of the pituitary body, it is obvious that laboratory experiments of this nature, performed on such small animals as puppies and kittens, for example, are beset with great difficulties in execution; and it would be natural to suppose, as Pershing has remarked, that just as a minute remnant of the thyroid left in an operation will prevent the development of myxedema, so a proportionate amount of the pituitary might escape detection and, remaining, defeat the purpose of the experiment.

Normal Pituitary in Alleged Acromegaly. — An analysis of the cases in which it has been averred that autopsies do not always reveal changes in the hypophysis tends rather to the casting of doubt on the diagnosis of acromegaly in such instances. Such cases must be very few. In Sarbo's case the patient was both tuberculous and syphilitic, and the jaw, nose and lips do not appear to have been enlarged. The skull was that of a case of true osteitis.

The cases of the Hagner brothers, described by Friedreich and Arnold, are doubtful cases also. The jaw, nose, lips and tongue were not enlarged. They had extensive and general enlargement of every bone in the body, due to osteitis. Marie regarded them as cases of hypertrophic pulmonary osteo-arthropathy.

In Bouardi's case the pituitary is reported normal.

Absence of the Pituitary Gland. — Boyce, in an autopsy on a case of phthisis, discovered that the pituitary gland was absent.

Thyroid Gland. — This has been considered previously (see page 13).

Thymus Gland. — This has also been considered previously (see page 15).

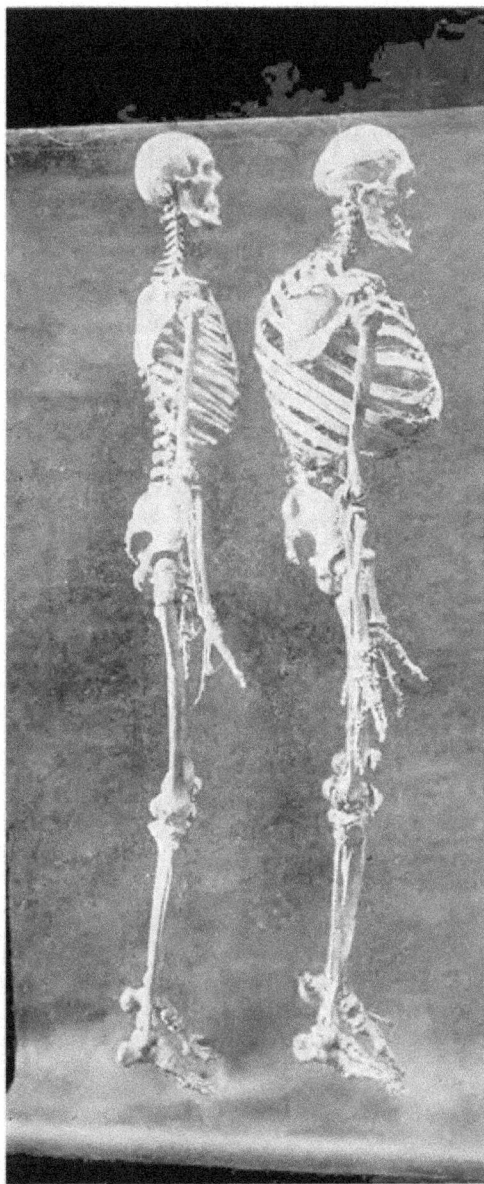

Fig. 23.—Normal and acromegalic skeletons (Osborne).

Changes in the Spinal Cord and Brain.—Arnold found the peripheral nerves and the lower part of the cord thickened, with increase of the interstitial connective tissue, with hyaline degeneration of the nervous elements. There was degeneration of the pyramidal tracts, and areas of cerebral softening.

Waldo has reported a case where an oval cavity, one centimeter in diameter, was found at the posterior extremity of the right hemisphere, and a second smaller cavity at the posterior part of the second temporo-sphenoidal convolution; also two others at the anterior portions of each of the lobes of the cerebellum. They were probably recent and due to emboli.

I have already referred to the coexistence of acromegaly and syringomyelia (see page 28).

Bone Structure.—The bone shows an undeniable dilatation of the vascular orifices and an enlargement of the grooves for the course of vessels along the bones (Marie, Broca). Klebs has noticed, in many portions of the affected extremities, the larger size of the arteries and capillaries of the skin, gaping, on section, and infiltrated and surrounded with young neoplastic tissue. This leaves conspicuous vascular canals and osteophytic growths in many of the skeletons, so that, judging from these preparations in museums, we notice strong resemblances to the changes in osteo-arthritis and, in a lesser degree, the appearances in osteitis deformans. We are in need of a good account of the minute bone changes in acromegaly.

Other Tissues.—Marinesco found that in the great toe all the structures were hypertrophied, even the membrana propria of the sweat and sebaceous glands, and that the external coat of the blood-vessels had undergone a notable hypertrophy. But most marked of all was the increased thickness of the sheaths of the subcutaneous nerves. The nerves themselves were degenerated; the alterations in larger nerves were much slighter. In the bones there was a new formation of bony tissue and an increase in both the length and thickness of the bones. The enlargement of the tongue was found to be due to increase in the connective tissue and the size of the bundles. The cervical sympathetic was found sclerosed, especially the lowest ganglion. The kidneys showed a condition of cortical nephritis. The follicles of the atrophied thyroid were enlarged and cystic, the nasal mucous membrance was thickened and mammillated, and the vessels were sclerosed and, in parts, almost obliterated.

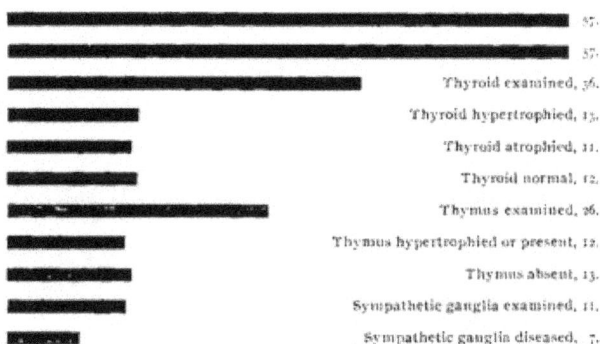

	57.
	57.
	Thyroid examined, 36.
	Thyroid hypertrophied, 13.
	Thyroid atrophied, 11.
	Thyroid normal, 12.
	Thymus examined, 26.
	Thymus hypertrophied or present, 12.
	Thymus absent, 13.
	Sympathetic ganglia examined, 11.
	Sympathetic ganglia diseased, 7.

FIG. 24.—Autopsies in 57 undoubted cases of acromegaly. Three doubtful cases are not included herewith. In each of the 57 cases the hypophysis was examined and found to be diseased.

ETIOLOGY.

During the years that have elapsed since the discovery of this remarkable disease, no great advances have been made as regards the true explanation of its causation. The various theories that have been enumerated will be duly considered hereafter.

Heredity.—It has been denied that heredity has any influence in the causation of acromegaly. There is, however, some evidence to show that it has been observed as a congenital affection.

Cenas has described the case of a boy aged 15 years which he believes to be congenital acromegaly. At birth there was asymmetry of the face and, after a few months, the child exhibited deformities of the hands. At four years there were pigmented spots on the hands and feet; these members had increased steadily in size at seven years, but there was only slight change in his arms and legs; they were clumsy. The head was enlarged and the hands were covered with violet discolorations. The feet were enormous. There was asymmetry of the tongue, cheek, and lips. The lower jaw was hypertrophied and there was kyphosis.

The famous cases of the Hagner brothers which have been studied by Erb and Arnold have been cited to support the view that a family type may exist. Marie, however, denied this, as regards these cases; but Jonathan Hutchinson gave his opinion regarding them as follows: "If I may venture a suggestion in a question so full of intricacy, it would be that they are, after all, examples of what may be called the family form of acromegaly."

It was Arnold who made the autopsy in one of these cases, and he gave to the affection, as he observed it in these subjects, the name of "pachyacria."

In Field's case, reported as acromegaly, although of doubtful type, the symptoms were present and the case published when the child was seventeen months old.

Moncorvo reported, in 1892, a case of what seems to be congenital acromegaly. It occurred in an idiotic child of fourteen months. The remarkable feature of the case was that the head was microcephalic, while the hands, fingers and feet, the nose and lips, were very much hypertrophied. Von Recklinghausen and Verstraeten believe in a hereditary influence in the etiology of the disease.

Sex.—In 130 cases in the appendix we find that seventy-three were males and fifty-seven females.

Race.—One case, that of Berkley, was in the negro race.

Age.—The youngest case reported is fourteen months. The oldest patient, at the time of report, was that of Gordinier, aged 77.

The age at onset of the disease was as follows: 0–10 years, 4 cases; 11–20 years, 19 cases; 21–30 years, 33 cases; 31–40 years, 22 cases; 41–50 years, 10 cases; 51–60 years, 2 cases; 61–70 years, 1 case; 71 years, 1 case.

Depressing emotions have been assigned a place in the etiology of the disease. I believe, however, that this is wholly an effect and not a cause of the disease. Chills have been mentioned as a cause, but I see no reason for this.

Trauma.—This has, no doubt, some influence in setting in operation the cerebral changes which are found in acromegaly. I have found trauma noted in the history of 19 out of 130 cases. Marinesco's case is a striking instance in point. The patient, who was 25 years of age at the time of her accident, fell from the second floor of a building and received an injury to the head from which she was unconscious for three hours. Epilepsy supervened, and also the usual symptoms of acromegaly. Baruch's case had a history of having fallen twenty feet, striking upon his head. This occurred five years before the patient's death and about two years before the symptoms of acromegaly were observed.

Gauthier's patient had two falls in which his head forcibly struck the ground. The patient attributed his disease to these accidents.

In Farge's case there was a severe accident, disabling the patient for ten months, and it was during this time that the first symptoms of acromegaly were said to have been noticed.

Other cases in which trauma has been noted are those of Unver-
richt, Barclay, Benson, Marie, Bourneville and Regnault, Bramwell,
Pel, Chauffard, Gauthier, Lavielle, Ruder and Marinesco (2),
Brooks, Strzeminski, Haskovec, and Remington.

Rheumatism is also said to be a cause. We have seen, however,
that in the erethic stage there are hyperesthesias and pains about
the extremities that resemble rheumatism very closely and that have
given rise to the belief that many of these patients had rheumatism.
Such was true of the cases of Vinke and Cunningham, and of
others.

Arthritis, as it is observed in the early stages of acromegaly
(*e.g.*, in the case of Middleton), is probably a purely trophic
change.

Infective Diseases.—These do not play a very conspicuous part.
Syphilis is not a factor in the histories of these cases; neither is
alcoholism, variola, scarlatina, or intermittent fever—although, naturally, the patients have had some of these affections at some time in
their past history.

THEORIES OF THE ETIOLOGY OF ACROMEGALY.

A review of the various theories which have been advanced to
explain the occurrence and course of acromegaly would lead us into
a maze of discussion from which we naturally shrink. The general drift of the later writers leads to the belief that we are dealing
with a disease of a trophic nature, depending on a perversion of
function of the pituitary gland; these changes being brought about
through the medium of an "internal secretion, like that of the
thyroid, which acts chiefly upon the trophic fibers of the peripheral
nervous system."

Although I shall endeavor to give, as briefly as may be,
the opinions of twelve competent men in the premises, it will be
found that the jury will fail to agree upon a verdict; for each one
takes a distinctly different view of the case, as it has been presented
for his judgment.

Marie's own view is entitled to the first consideration. His is
the "*theorie hypophysaire,*" in which he attributes to the changes in
the pituitary body all the phenomena of the disease. He believes
that this gland possesses a function analogous to that performed by
the thyroid body in myxedema. The changes throughout the
system constitute a systematized dystrophy, depending on an alteration of the hypophysis, which henceforth becomes incompetent to
neutralize certain toxic substances the accumulation of which in the

blood brings about the various changes of acromegaly. Marinesco supports this position.

The Theory of Trophic Neurosis. — In this theory, not yet demonstrable by any anatomical evidence as far as I can find, the nervous system is held accountable, primarily, for all the pathological phenomena. The changes observed in the pituitary body are, therefore, considered to be of a secondary nature, this structure sharing with other organs the morbid process.

The added influence of the diseased pituitary, which we have seen to be, almost without exception, disordered, will be felt in the phenomena of autointoxication and, hand in hand with the action of the diseased thyroid body, the trophic disturbances are augmented. This is the theory of von Recklinghausen and Holschewnikow, and is supported by Mossé of Toulouse. Mossé cites in its support the early occurrence of vasomotor disturbances, local asphyxia of the extremities.

Thus we have, first, a trophic neurosis; second, toxic phenomena, due to disorder or suppression of the internal secretions of the pituitary and thyroid glands. The author believes that this is the more rational view, and that in the absence of anatomical proof the more favorable results following the administration of thyroid extract, either with or without the use of the pituitary gland, afford a therapeutic corroboration of this line of argument. The use of these preparations is based on the belief that they supply the protecting elements which, in its diseased state, the system lacks, and that they are able to check, to some extent, the toxic process. This theory is in harmony with that of Horsley, who seeks in a similar way to establish the relation of the thyroid body to myxedema. That the pituitary and thyroid bodies and the thymus are interrelated seems highly probable, when we consider the striking analogies between the dystrophies of acromegaly, myxedema, and cretinism, and the changes in all three of these glands observed in the various forms of acromegaly.[1]

It may be that in case of impairment, destruction or physiological quiescence of either of these glands, the remaining glands may fulfil its function, just as, after removal of the spleen, life is preserved by compensatory action (see Kron's case).

Claus' investigations, as well as those of Rogowitsch, Vassali, Sacchi and Hofmeister, seem to show that ablation of the thyroid was followed by hypertrophy of the hypophysis, and *vice versa.*

[1] See monograph by G. Buschan on Myxedema and Allied Conditions and the Physiology and Therapeutic Uses of the Thyroid Gland, Leipsic, 1896. Extensive bibliography; 182 pages.

According to Schaefer and Oliver, they are physiologically antagonistic. In view of the superficial resemblance, in some respects, between acromegaly and myxedema, it might be held that "acromegaly is a result of a disturbance of a chemico-physiological equilibrium, maintained, in health, by the normal activity of these two glands . . . and thus the cause of the disease is perverted, and not suppressed or exaggerated, functional activity of the gland" (Rolleston).

Vassali and Sacchi, by their investigations, show that the function of the pituitary is essentially trophic, enabling the nerve tissues to take up and assimilate oxygen from the blood-stream. It also exerts an influence on metabolism, destroying or rendering innocuous certain waste products.

According to Tamburini, there are two distinct phases. In the first, there is hyperfunction of the gland and hypertrophy; in the second, there would be an adenomatous change, as in the case of Claus and Vanderstrict, or sarcomatous, as in Dallemagne's first case, or cystic, as in his second and third cases; and then, clinically, would follow cachexia and death.

Massolongo supports Tamburini and assigns acromegaly to the class of autointoxications. He seems to have antedated, by two years, Tamburini's views as to the hyperfunction or hyperactivity of the pituitary.

Dallemagne shows that we have not yet arrived at even an elementary knowledge of the laws governing the equilibrium of nutrition. We must admit the fact of such an equilibrium; the proportions of our various members, the dimensions of organs, and our entire anatomical type are governed by natural laws, the operation of which we have not even begun to understand.

It is highly probable that nutrition or trophic centers exist at various points in the central nervous system. Dallemagne is inclined to connect the changes which the pituitary body undergoes with "*proliferations qui encombrent le canal central,*" which latter is considered as, after all, simply the upper cul-de-sac of this canal.

On the supposition that the central canal is emunctory in its function, its epithelium would have a rôle analogous to that of the renal epithelium, and we may assign at various points in the cord these centers of equilibrium of nutrition, to create a sort of nutritive metamerism.

Thus it would follow that disturbances of these centers of equilibrium would be followed by exaggeration of form, due to the anomalous state of nutrition, and, further, abnormal products of this diseased nutrition would be put in circulation. In this manner irritative troubles of a secondary nature would be set on foot and, perhaps, particularly about the ependymal (central) canal. Thence would arise the affections assigned to this canal, and especially the modifications of the hypophysis.

Dallemagne holds that the changes observed in the hypophysis are a result and not a cause, no matter how the cause operates. The origin of the troubles must be in the centers which preside over equilibrium of nutrition, on which the proportion of our various members depends.

As regards bone formation, the views of Marinesco are of interest in this connection. He holds that the development of normal bone is dependent upon the pituitary and thyroid glands. "It is probable that they secrete substances, such as soluble ferments, necessary for stimulating and carrying on the normal osteo-genetic process." We know, moreover, as Hofmeister has shown, that there exists normally in thyroidectomized animals an atrophy of the osseous tissue. Or else, in acromegaly, the equilibrium of nutrition of bone is disturbed and the nerve force necessary for the regulation of the osteo-genetic process is destroyed; then is produced this abnormal process of the bones of the extremities and of the bony extremities.

Other theories as to the pathology of acromegaly have a historic rather than a practical interest. We have the theory of Klebs. He assigns the affection to an exaggerated development of the vascular system, in which the thymus gland plays an important part. The affection would thus be a thymic angiomatosis, in which occurs a general proliferation of vascular germs or angioblasts. Gauthier seems to believe in the theory of angiomatosis.

Freund and Verstraetem refer this affection to a pathologic process, little known up to the present, but seated in the essential organs of reproduction. They discover in it the result of an "*inversion dans l'évolution de la vie génitale.*" According to their view, acromegaly belongs to a fresh pathologic family, originating in perversion of the genital life and including infantilism and gigantism. The genital function is, as a rule, found to decline at an early age.

This theory is based, to begin with, on the alleged appearance of the first symptoms about the age of puberty. This is too slender a foundation, in my observation. It is also based on the usual existence and early appearance of genital disturbances.

Freund is credited with an observation to this effect: that, in his studies of development, as seen in negroes and the anthropoid apes, he has been led to think that acromegaly is a sign of atavism !

Bacteriology has been invoked by Arnold to yield an explanation, but no information has been elicited from that source.

Without lingering further over the various speculations, we may group the views of investigators with reference to the function of the pituitary gland and its relation to acromegaly as follows:

That the essential symptoms are due to an alteration of the internal secretion of the gland. Three forms of modification are suggested (see Rolleston):

I. Suppression, a condition of incompetency of the pituitary gland, analogous to myxedema or Addison's disease. Such incompetency would imply either an atrophy of the pituitary gland, which is not in accordance with the post-mortem findings, the reverse being the case; or else destruction by a tumor.

It has been suggested that the simple absence of the internal secretion cannot be the entire explanation, because cases have not improved satisfactorily by treatment by pituitary extract. This argument, however, I consider to be open to criticism.

II. Hypersecretion, due to excessive activity. This view is borne out by observations of simple hypertrophy of the hypophysis. The administration of pituitary extract would be expected to aggravate the symptoms.

III. Perverted function, due to heterogeneous transformation of the structure of the pituitary body. Rolleston's statement on this head is clear, and we will give it as he makes it—taking it from Marie and Marinesco:

1. The secretion might be deficient in an active principle, the function of which is to restrain exuberant growth in the extremities. This, perhaps, is rendered improbable by the general failure of pituitary extract when given medicinally, and there is, of course, no evidence that irregular growth is spontaneously likely to occur if not kept in check.

2. The abnormal secretion might contain some toxic principle which stimulated the tissues to overgrowth. That the disease is a toxemia receives some support from its points of resemblance to hypertrophic pulmonary osteo-arthropathy, for Marie has suggested that the changes in the limbs characterizing the latter disease were due to the local action of poisons absorbed from the pulmonary lesions.

3. It might be thought that, owing to alteration in the secretion of the pituitary, the equilibrium and interaction between it and other internal secretions are disturbed. Possibly some toxic body, such as was suggested in the preceding alternative, might be thus produced.

It is possible, also, to suppose that there is some primary nervous lesion of unknown character and that the glandular changes constitute a series of phenomena of the second order. Arnold, Dercum and Dreschfeld have suggested such a view.

Cases occur occasionally which seem to be examples of general involvement of the glands of internal secretion, the inference being, very naturally, that these glands are affected by a cause common to them all. In five such cases collected by Murray acromegaly and exophthalmic goitre coexisted in each; in four of them glycosuria was present. In two fatal cases the thyroids as well as pituitaries were enlarged, and one of them showed an enlarged, persistent thymus gland. Lancereaux has called attention to similar cases.

ACROMEGALY AND GIGANTISM.

When it became known, five years ago, by the independent observation of Tamburini, Brissaud, and Taruffi, that the pituitary body undergoes a remarkable hypertrophy in the case of a large percentage of giants and was probably an agent of great importance in producing these wonders of the world, the hypophysis may be considered to have reached the pinnacle of its power. In the giants that came to post-mortem examination, the pituitary was found truly hypertrophied.

In the skeletons of giants that were forthwith examined, unmistakable evidences of acromegaly were found. The skeleton of which I furnish the measurements shows this in a moderate degree. The pituitary fossa is the largest on record, measuring 27 millimeters antero-posteriorly, 42 millimeters laterally, and 12 millimeters in depth. Thus we find that the celebrated "Irish giant," Cornelius McGrath, described by Cunningham, had undoubted evidences of acromegaly, including a huge pituitary fossa. So, also, it was found by Taruffi that the giant's skeleton which he examined in Reggio was also that of an acromegalic. W. Hutchinson, who has made a study of this subject, estimates that from 40 to 60 per cent. of giants are also cases of acromegaly. He cites the fact that of over one hundred cases on record, only one lived to old age, very few to middle life, and a majority died before the age of thirty, many deaths being due to a general failure of the vital powers, or some trivial intercurrent disease imposed upon this.

"Myth and popular impression to the contrary notwithstanding, giants are a short-lived, feeble-minded, weak-bodied race, shambling in their movements, and of the mildest dispositions. They are occasionally possessed of great power, but this seldom lasts more than a few years, just as is also seen in some cases of acromegaly. The position which appears to us best to harmonize these facts is that acromegaly and gigantism are similar, if not identical, disturbances of pituitary metabolism, the one beginning in early life, before the full stature has been reached, and producing comparatively symmetrical results, extending over a considerable period; the other developing after maturity and expending its overgrowth force at the points of least resistance to growth—the hands and feet, nose and lower jaw. In both, the essential process is a more or less rapid overgrowth, reaching a definite limit, and soon followed by correspondingly rapid decay. The ultimate result is the same in both: the situation of the outgrowth (extremities and distal portions of appendicular skeleton generally) is strikingly similar; there is the same disturbance of the sexual functions—in fact, almost the only differences between them are in the pressure symptoms (headache, hemianopsia) and more rapid course of the adult form.''

Maximilian Sternberg distinguishes two forms of gigantism—first the normal, and second the pathological, which almost always is associated with acromegaly. In thirty-four cases of gigantism examined, he finds evidences of acromegaly in fourteen (42 per cent.), which shows that gigantism predisposes to acromegaly; on the other hand, acromegaly predisposes less to gigantism.

It is in this second class that we find prognathism and enlargement of the sella turcica. Sternberg exhibited recently, in a discussion of this subject, a typical skull from both varieties of gigantism. He considers gigantism and acromegaly as separate entities and states that only about 20 per cent. of acromegalics are above 1.80 meters (6 feet) in height, and that above 40 per cent. of giants are acromegalic. Gigantism, therefore, is considered by him as predisposing to acromegaly; one of their tendencies to trophic organic changes being acromegaly. Nearly half the giants die from this cause. Marie puts it this way: ''Acromegaly is gigantism of the adult; gigantism is acromegaly of the adolescent.''

As to the two types of acromegaly, there is a disposition to assume the long or giant type of acromegaly, if the disease originated in the period of adolescence; but if the onset is delayed until later life, the type will be large (Brissaud and Meige).

In a case of Dallemagne acromegalic symptoms appeared fifteen years previous to observation, and for many years before that the patient had gigantic stature. After death the heart, kidneys, liver and spleen were double and even triple the normal weight. This author considers that this enormous increase should be attributed more to the acromegaly than to the gigantism. He terms it a sort of splanchnomegaly, referable to the same process as obtains in hypertrophy of the extremities.

A case described by Brissaud and Meige was as follows: The man presented nothing abnormal before the age of 16; he then rapidly grew larger, and at 21 years he measured 7 feet 2 inches and weighed 340 pounds. He remained well and spry until the age of 37, when his spine gave way under a strain and kyphosis followed, so that his nipples were on a level with the anterior superior spines of the ilium. Mental dulness, debility, fatigue, bronchitis, night sweats and headache harassed him. The bones of the face and extremities were characteristically enlarged. Circumference at nipples, 62 inches, and over the most prominent part of the kyphosis and pigeon breast, 74 inches.

Prof. Woods Hutchinson has published an interesting record of an autopsy in a giantess. The height of this individual was 6 feet 7¾ inches. She was a native of France and died at the age of 17 in the United States quite suddenly, after a short attack of grippe or bronchitis, her illness lasting only three days, but having been preceded by several years of failing strength. She was greatly emaciated, and this fact led to the general belief that she had died of consumption.

At the autopsy a remarkable condition of affairs was observed. The great height was found to be due to greatly enlarged extremities, as compared with the trunk. The hand was 11.25 inches in length, or nearly one-seventh of the height. The foot was 13.75 inches long. The lower jaw measured 6.25 inches from angle to symphysis, as against 3.75, which is the normal for male adults. The skull itself, however, measured 21⅛ inches in circumference, as compared with the normal measurement (female) of 20 inches.

The genital organs were deficient; the mammary glands were almost completely absent. The mons veneris and labia were flat and poorly developed. The vagina was small and straight, barely admitting the forefinger. The uterus was 1.25 inches long and ⅔ inch broad, about the size and shape of the last joint of the little finger, and weighed two drachms. The Fallopian tubes were barely recognizable. The ovaries were small, granular-looking masses, about the size of the finger-nail, adhering to the surface of the broad ligament. The skeleton was spongy and crumbling, a good example of osteo-porosis.

The skull showed enormous frontal sinuses and a huge pituitary fossa. The latter measured 1.25 inches in antero-posterior diameter, and 1.50 inches transversely.

The author has examined a giant now living, named Henry Alexander Cooper, who was born in the county of York, England, on the 12th of March, 1860. He is seven feet and five inches in height with boots on, and the circumference of his head is twenty-five inches. There is no prognathism, no exophthalmus, and no hemianopsia. There was formerly severe headache, and at one time pain in his eyes with some blurring of vision at about the period of greatest growth. The brows are rather prominent, the nose large; the chin, however, is not unduly prominent, considering the size of the face. The latter is of oval shape. The hands and feet are not disproportionate to the size of the skeleton. The abdomen is not prominent, and there is neither excessive thirst nor hunger. The man seems to be losing flesh. It was not possible to ascertain his exact weight. This giant is not, therefore, a case of acromegaly. It should be stated that this man's intellect is perfectly clear, his manner pleasing, and his voice agreeable.

I cannot leave this subject without referring to the masterly description by Prof. D. J. Cunningham of the Irish giant, Cornelius McGrath. This individual was born in Ireland in 1736 and died in 1760, being 23 years old. His height was seven feet two and one-quarter inches, as measured by the skeleton now in the Anatomical and Zoological Museum in Dublin. This, however, is less than the height of the skeleton I describe and less than that of Charles Byrne's skeleton, now in the Royal College of Surgeons of England, viz., seven feet seven inches.[1] The skull of the Irish giant has a cubic capacity of 1600 cubic centimeters; that of Byrne, 1520 cubic centimeters; while mine has a capacity of 2320 cubic centimeters. I will, however, reserve comparisons until I give the description of my case in full.

It may be said before closing that Virchow is generally quoted as believing that there is no connection whatever between the partial giant growth which is seen in ordinary cases of acromegaly and general giant growth, but most authorities, including Langer, Cunningham, and W. Hutchinson, unite in holding an opposite position.

MIXED PREMATURE AND IMMATURE DEVELOPMENT.

Cases have arisen which have suggested that there is, paradoxical as it may seem, a relation between gigantism and dwarfism. Such cases are those described by Mr. Jonathan Hutchinson in 1866 and by Mr. H. Gilford in 1896. In the latter case, while the patient was clearly a dwarf, there were parts that were more than fully

[1] Comparative measurements of these and other skeletons will be given later.

developed; and Mr. Gilford was led by this case to the study of dwarfism and gigantism. He sees a close relationship between these deviations in nutrition, and suggests the term micromegaly as descriptive of his case and others allied to it. He thinks it not impossible that the cause of acromegaly operating before birth may bring about micromegaly; for many giants have evidently owed their proportions to the former. May the one be the congenital condition of the other, or are the two opposite states?

DIAGNOSIS.

In a large proportion of cases the diagnosis of acromegaly is made without difficulty. The symptoms are so characteristic and, once seen and recognized, are so indelibly impressed on the observer, that subsequent cases are not readily overlooked when the affection is well established. It is obviously important, however, to recognize these cases at the outset, and if medication will accomplish anything, as we believe it may, the task of counteracting the disease will be the easier.

Without repeating what has already been said in the preceding chapter on symptomatology of typical cases, I will reserve my remarks for the atypical forms and for some of the diseases with which the affection may be confused. As new diseases become better known, it becomes possible to detect and classify varieties typical and atypical, or, as the French term the latter class, "*les formes frustes.*" Many refinements of diagnosis have been made in the case of other affections; thus we may recognize exophthalmic goitre without goitre; tabes may be diagnosticated while latent and without incoordination; and Chauffard therefore asks why we may not diagnosticate acromegaly without hypertrophy of the members. The difficulty probably would be that it is the members that present the first objective symptoms. But we shall probably hear that diagnosis will be made on the strength of ocular changes, paresthesiæ, headaches, genital disorders and deficiencies, and hypertrophies at the minor terminal points in the body.

Four years after Marie discovered acromegaly, he was able to differentiate another affection which bears a very strong resemblance to it, but which appears to be a separate entity. Marie says of this affection: "Sous l'influence de micro-organismes, la production au niveau des lésions de l'appareil respiratoire de substances purulentes ou fermentées, passant en suite dans la circulation, exerce une action elective sur certaines parties des os et des articulations, pour determiner les lésions de l'osteo-arthropathie hypertrophiante."

The following parallel by Rauzier gives the differential diagnosis of the two affections:

ACROMEGALY.	PULMONARY HYPERTROPHIC OSTEOARTHRO-PATHY.
1. Large stumpy hands; fingers uniformly hypertrophied, with preservation of the proportions of the phalanges; nails small and flattened.	1. Hands deformed; fingers greatly enlarged, club-shaped tips; nails enlarged, lengthened and curved, striated and cracked. Hyperextension of the phalanges possible.
2. Great increase in size of the carpo-metacarpal region.	2. Slight hypertrophy of the heads of the metacarpals; nearly normal size of the carpo-metacarpal region.
3. Hypertrophy of the wrist, proportionate to that of the hand, without deformity.	3. Great deformity of the wrist, which is enlarged, swollen, and forms a considerable prominence above the dorsum of the hand.
4. Same degree of change in the feet as in the hands.	
5. The lesions are in the bones and soft tissues alike.	5. Alterations affect only the bones, particularly the epiphysis, in certain joints. The soft parts do not partake in the enlargement affecting the bones. Slight edema.
6. Kyphosis, common symptom — cervico-dorsal.	6. Kyphosis not so frequent and late in onset. Scoliosis frequent.
7. Great hypertrophy and exaggeration of the curves of the lower jaw.	7. Lower maxilla normal. Occasional thickening of the alveolar border of the superior maxilla.
8. Obscure pathology; probably defect of nutrition.	8. Pulmonary origin.

We recognize the same agencies in the production of the well known "club-fingers" of tuberculosis. The circulation carries the microbes and their products to the extremities of the fingers, where they react on the bones and surrounding tissues to produce hypertrophy.

The question has been raised whether hypertrophic pulmonary osteoarthropathy is a separate disease or a transition stage between the rheumatic arthropathies and acromegaly. We are not in a position to determine all these inter-relations; but no doubt further observations will soon clear up the different dystrophies of this class, if we may judge by the great advances made in the last decade.

Myxedema.—We have here, as in acromegaly, enlargement of the face and limbs, with the difference that in myxedema the hypertrophy affects the soft tissues exclusively; the skin is greatly thickened, of a brawny nature and a yellowish or waxy color, and is not movable on the underlying structures; while in acromegaly the hypertrophy affects all the tissues and the skin is more flexible and movable. It is customary to speak of the face of myxedema resembling the full moon, because of its puffiness. The face in acromegaly, on the other hand, is elongated. In myxedema the thyroid is usually atrophied, mental weakness is common, bodily stature is small, and the disease is more common in women than in men. There is no prognathism, kyphosis or symptoms referable to an enlarged pituitary body. (See Dr. Norman Dalton's case of combined myxedema and acromegaly.)

Osteitis Deformans (Paget's Disease).—This is accompanied by curvature of the long bones of the skeleton, especially the femora and tibiæ. The skull is broad at the top and the entire contour of the head is, in general, that of a triangle with the base upward. In myxedema the face is round, while in acromegaly the face shows a distinct oval. Osteitis deformans has never been known in patients under forty years of age.

In acromegaly, as Marie puts it, the osseous changes are seen *in the extremities of the bones and the bones of the extremities*. In Paget's disease, on the other hand, the shafts of the bones show characteristic changes, such as bowing of the tibiæ and femora, or even of the humerus and forearm. In osteitis deformans the cranial bones are chiefly affected, while in acromegaly the facial bones undergo hypertrophy.

The skeletal changes in acromegaly tend to take place symmetrically, while in osteitis deformans a single bone, such as the tibia or femur, may be first attacked, and after a considerable interval the corresponding bone of the opposite side presents similar changes; but the bone originally involved will usually present a greater hypertrophy than the bone secondarily involved.

Elephantiasis.—This is also usually an asymmetrical affection. The hypertrophy involves the entire member. The lower jaw and spine are not involved. The skin is indurated, very thick, dry, and difficult to move. Ocular symptoms are, of course, absent. (See Lombroso's case and the case of the Hagner brothers, described by Arnold, for elephantiasis.)

Chronic Rheumatism.—This has been the diagnosis in many cases seen in the early stages. In the case of giants that have had acromegaly, rheumatic pains are noted and also what were naturally termed "growing pains." The "Irish Giant" was subjected to baths in salt water for symptoms of this nature. The pains of rheumatism are chiefly in the larger joints and are accompanied often by grating and unilateral wasting. The face escapes. There is usually a history of an acute attack, and cardiac symptoms ensue.

It has been remarked that disturbances simulating acromegaly may also occur in the course of tabes. Partial acromegaly, or more properly partial hypertrophies, occur. They are, as far as I have seen, congenital, involving one side, and stationary. They are variously designated as macrodactyly, macropody, hypertrophy of a single member or of one-half of the body. If they have anything to do with acromegaly, we must look for it in some prenatal condition not yet understood.

This latter-day means of diagnosis has been used by Marinesco, who has submitted the hands of four subjects of acromegaly to Roentgen photography, as follows: Three of the patients belonged to the massive type described by Marie, while the fourth, a woman of 33, was of the gigantic type. In the first case, a man aged 55, of the massive type, whose affection was of eleven years' standing, the skiagraphs showed a marked hypertrophy of the metacarpal bones and phalanges. This hypertrophy was uniform; it was merely an exaggeration of the normal state. The epiphyses and apophyses of the last phalanx were hypertrophied and measurements were readily made from the skiagraph. In the third patient, aged 33 years, who presented the giant form of the disease, of eight years' duration, the skiagraph showed that the hypertrophy of the soft parts was much less than in the first case. In the fourth patient, aged 33, an example of the massive type which was of at least seven years' duration, the skiagraph revealed a hypertrophy of the extremities of the phalanges and of the metacarpals. In this case the distance from the metacarpo-phalangeal articulation to the extremity of the last phalanx was 11 centimeters 2 millimeters. There was hypertrophy in length and in thickness. It was thus easy to demonstrate the clinical type to which a case belongs by noting the relation between the hypertrophy of the soft parts and the bones. In acromegaly of the massive type, the hypertrophy of the soft parts is much greater than in the case of acromegaly of the giant type. Marinesco also examined a case of erythromelalgia in the same manner, and failed to find any lesions of the skeleton, which shows that this affection has no relation to acromegaly.

In Sir William Broadbent's case the skiagraphs showed that the hypertrophy was wholly in the soft parts.

TREATMENT.

The treatment of acromegaly is more encouraging than the brief references to it in the text-books would lead us to suppose. We understand, of course, that the palliative or symptomatic treatment is of great importance in improving the general comfort of the patient. The intense headaches, the paresthesiæ, the pains, which

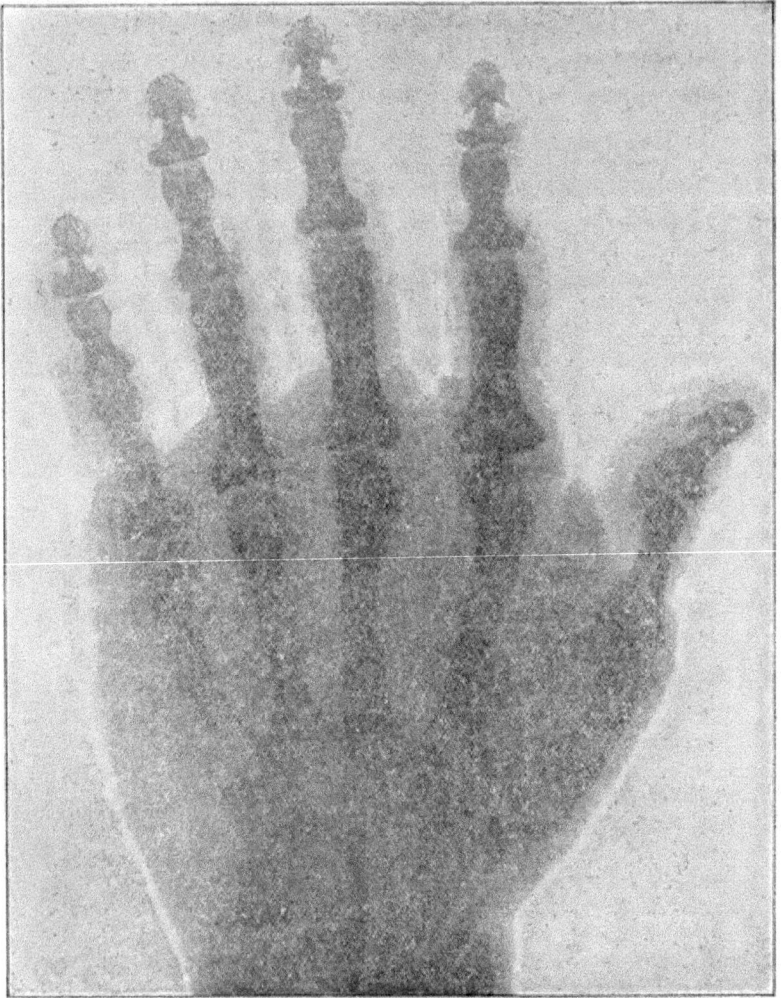

FIG. 25.—Skiagraph of right hand from case of acromegaly (Osborne).

in many cases have been considered rheumatic, and the general depression of spirits, may be treated with more or less success by various drugs and measures directed to the nervous system. Chief among these we may mention phenacetine, antipyrin, iodides, bromides, and caffeine. Headaches that are refractory have been greatly mitigated by the use of cannabis indica and acetanilid. The pains in the extremities are sometimes benefited by the use of codeine or morphine. Iron, arsenic and strychnine have their place in giving tone to the enfeebled muscles of the later stages.

Glycosuria, if it be present, also demands attention and should be treated by a suitable diet, by the use of salicylates, alkalies, codeine, or other drugs suited to that particular condition.

It is, however, in the use of animal extracts that the general treatment of the disease will probably receive the greatest benefit. While much discussion has attended the explanation of the action of thyroid extract and pituitary extract, the fact remains that these agents are being used more and more in diseases of the nervous system, and in the case of acromegaly thyroid extract seems to be the favorite. It is remarkable, however, that so many cases elaborately recorded in all other respects are destitute of any note whatever as to treatment.

Brown-Sequard was the first to advocate the use of thyroid and other animal extracts in acromegaly. In a communication to La Société de Biologie on May 20, 1893, he said: "Ce sont les liquides retires de la rate de la thyroide et de la moelle des os, qui paraissent certainement devoir posséder le plus de puissance contre cette terrible maladie."

Although much ridicule was indulged in by the medical profession and the laity when Brown-Sequard first enunciated his belief in the rationale and efficacy of extracts made from various animal tissues, belief in their use is steadily gaining ground. Thyroid extract, in particular, has assumed what seems to be a permanent place in the treatment of various diseases, notably myxedema, and its use in obesity, mental troubles and other affections is rapidly spreading. The use of hypophysis cerebri is not mentioned by M. Bra in his elaborate work on animal extracts, and as far as I know was not mentioned by Brown-Sequard. The latter employed the hypodermic method, and his plan involved the use of "injections quotidienne et, a parts egale, de liquide de rate et de moelle des os, et d'adjoindre au besoin a ces deux liquides, le liquide orchitique." I do not know of any cases in which this plan of therapy has been carried out.

Cases have been treated with thyroid extracts by the following authors: Bramwell, Benson, Sanger Brown, Bruns, S. Solis-Cohen, Costanzo, Eschle, Eshner, Jeffrey, Mendel, Mossé, Middleton, Murray, R. N. Parsons, J. J. Putnam, Schultze, Unna and Vinke, Sears, Comini, Witmer.

Bruns' patient was a woman, aged 24, who, after an abortion in her first pregnancy, experienced vague pains and forms of paresthesia in the hands and feet, which were noticed to become gradually enlarged. Two years later she presented a typical picture of acromegaly, with enlargement of bones and soft parts of the face; marked thickening of bones of trunk, especially the clavicles; hands and feet enormously enlarged; patient extremely nervous and excitable, sleeping very badly; suffering from almost constant headache; vague pains and abnormal sensations in extremities, preventing her from doing any fine work. Tabloids of thyroid extract (quantity unknown) were given in increasing doses until four were taken per diem. The enlargement of parts persisted, but the improvement in subjective symptoms was striking; the nervous excitability rapidly diminished, the patient slept well, and the headache disappeared; the pains in the extremities also vanished, and the patient was able to resume the finest work.

Benson's and Mossé's cases treated with thyroid extract also gave good results. In the latter the chief benefit was observed in the mental symptoms. The patient afterwards died, and a large sarcoma of the hypophysis, weighing thirty - six grammes, was found.

In a case recorded by J. J. Putnam, thyroid powder in doses of fifteen grains every second day made great improvement. The hands and feet diminished steadily in size, so that the patient (a woman) at the end of three weeks was able to resume work with her hands, and after further treatment could "wear boots that she could not get on at all when her symptoms were at their height."

In Feliciano Costanzo's case remarkable improvement followed the use of the thyroid extract for a period of eighty - one days, including a cessation of two weeks on account of physiological symptoms due to the drug (pulse 140, vomiting, temperature 39–40° C., pain in the head and limbs). The dose was three teaspoonfuls administered by the mouth three times a day.

	Before treatment.	After treatment.
Right hand—Circumference of thumb	7.5 cm.	7 cm.
Circumference of index finger—		
First phalanx	7.5 cm.	7 cm.
Second phalanx	7.5 cm.	5 cm.
Third phalanx	6.2 cm.	6 cm.

	Before treatment.	After treatment
Circumference of middle finger —		
First phalanx	8 cm.	7 cm.
Second phalanx	7.5 cm.	6 cm.
Third phalanx	6 cm.	5.5 cm.
Left foot—Circumference of great toe	12.5 cm.	9.5 cm.
Circumference of second toe	7.5 cm.	6 cm.
Circumference of third toe	7.5 cm.	6 cm.
Circumference of fourth toe	8 cm.	6.5 cm.
Circumference of fifth toe	7 cm.	6 cm.
Circumference of left foot	26 cm.	24 cm.
Body weight	67 kg.	59 kg.

It is thus evident that to obtain the best results thyroid extract should be pushed to its physiologic and even toxic effect.

Vinke treated his patient, a woman of 51 years, with both thyroid extract and pituitary extract, as follows: During the first two months she took one-half grain of thyroid extract three times a day. One grain of this extract represents ten grains of the fresh gland. During the third and fourth months, desiccated pituitary bodies were given, about one and a half grains three times a day. During the last month a combination of both extracts was administered, four grains of desiccated pituitary bodies and one-half grain of thyroid extract a day. By comparing the measurements at the beginning of the treatment and five months later, it will at once be seen that there is a decided decrease in size.

Pituitary extract has been used by the following: D'Esterre, Bard, Bramwell, Broadbent, Dodgson, Marinesco, Mendel, Murray, Ransom, Rolleston and Vinke, Eshner, Dercum, Thomas, Schultze, and Osler.

Marinesco has used pituitary extract in three cases. The first two, a woman of 53 years and a man of 54, belonged to the massive type; the third was a woman of 30 years, of the gigantic type. In the first case there were such violent headaches that suicide was contemplated. The pain was soon relieved, but nausea set in, and on an increase of the remedy there was colic and diarrhea. There was improvement while taking four tabloids daily, but before long the headaches returned. The variations of the pulse were not such that a definite conclusion could be drawn from its influence on the circulation. The urine was increased from a liter to a liter and a half and even two liters daily. No change was noted in the size of the limbs.

A second case, treated by Marinesco with pituitary extract in doses gradually increased from one-half a tabloid to four tabloids daily, gave at first no especial symptoms, subjective or objective; but a gradual relief from headache followed in two weeks. The pains in the hands became less and their movements more free. In two months it was possible to say that the soft parts involved were less swollen than before the treatment was instituted. The headaches, the paresthesias in the hands and the pains in the lower members were lessened. The pulse varied from 60 to 80, and the urine reached 1300 grammes.

In Marinesco's third case, a diabetic, the urine was increased quickly from sixteen liters to twenty-one liters. The bodily weight at first declined and then increased.

Sir W. Broadbent treated one case with pituitary extract with some improvement of mental condition, although in this case it was believed that the improvement may have been referable to the relief of the excessive constipation.

It is probable that a combination of pituitary and thyroid extracts would be of service in cases where both of these structures are diseased. It is not out of place to cite, in this connection, the remarkable effect which thyroid extract produces when given in the treatment of obesity. Out of 145 recorded cases it produced a decided effect in 96 per cent.; the loss of weight being in respect to fat, watery constituents, and albuminoids. Thus we see that overgrowth, entirely independent, as far as we know, of any change in the hypophysis, also responds to this medication.

NAME.	AMOUNT.	RESULT—PITUITARY TREATMENT.
Bard's case		No result; rabbit's gland, subcutaneously.
Bramwell's		Improved; thyroid treatment failed.
Bramwell's (2d)		Unimproved; thyroid treatment improved.
Broadbent & Dodgson's		Improved mentally.
Dercum's		Unimproved.
D'Esterre's	2 grs. t. d.	Lost eight pounds.
Eshner's		Unimproved.
Marinesco's (1st)	4 tabloids daily.	Greatly improved.
Marinesco's (2d)	½ to 4 tabloids daily.	Greatly improved.
Marinesco's (3d)		Slightly improved.
Mendel's	15 grs. daily.	Improved; failed to improve under thyroid.
Murray's		
Osler's		Unimproved.
Ransom's		
Rolleston's (1st)	5 grs. b. d.	Unimproved; used thyroid extract simultaneously.
Rolleston's (2d)	5 grs. b. d.	Improved as to headache; used thyroid extract simultaneously.
Schultze's		Worse; heart disturbed.
Thomas's		No result; effect unpleasant.
Vinke's	1 to 4 grs. t. d.	Greatly improved; alone and combined with thyroid.
Witmer's	4 grs. t. d.	Unimproved at two months; slightly better at three months.

French writers have questioned how much the influence of suggestion may have to do with the beneficial results noted in cases of nervous diseases treated with the organic extracts. It is naturally difficult to settle such questions conclusively.

Marinesco formulates two propositions:

1. Tablets of pituitary gland exert in acromegaly an "elective action," particularly upon the cells of the tumor, which have preserved their integrity.

2. They act upon the intracranial pressure, or rather upon the vessels of the pituitary tumor.

The observed effects might be explained by either proposition. The remedy may be able, when introduced into the circulation, to excite the diseased hypophysis or aid its action and thus bring about a lessening of the disturbances.

Or, on the other hand, which the experiments of E. Schaefer and Oliver tend to prove, the extract of pituitary gland has a vasomotor action by which the constriction of the vessels of the highly vascular pituitary tumor would be able to exert a contraction of that body, and consequently a modification of the intracranial tension, a process which would readily explain the relief from headache. They found a rapid and considerable rise of blood-pressure, due to a direct action on the arterioles and probably on the heart muscle.

These questions are still unsettled, and we remain in ignorance of the mechanism of the production of acromegaly.

E. Schaefer has shown that the action of thyroid extract is to diminish blood-pressure, while pituitary extract greatly increases the pressure, as I have stated.

Isaac Ott has made a contribution to this subject in a "Note on Animal Extracts," published in 1896. His experiments were made upon rabbits, and he found, as did Schaefer and Oliver, that the thyroid lowered the blood-pressure, but with this difference: that it lowered the heart beats, even when he used cold filtered infusion of thyroid. This lowering of the heart beat ensued when the vagi were cut or when their peripheral ends were paralyzed by atropine. Thus it is probable that the drug acts upon the heart itself and thus lowers the cardiac beat. The fall of pressure is not due to a paralysis of the main vasomotor center, for the arterial tension falls quite as readily when the cord is cut. The cold filtered infusion of the pituitary gland produced a rise of blood-pressure and a fall of pulse. The subcutaneous injection of pituitary powder caused hardly any rise of temperature.

Szymonowicz, in two experiments, obtained a slight fall of pressure and a quickened heart beat. Mairet and Bosc noted very slight physiological effects from the use of pituitary extract beyond a slight rise of temperature and depression with some digestive disturbance.

Prof. W. H. Howell has injected into the circulation of dogs extracts of the hypophysis of sheep. They were made in normal saline, or in glycerin, followed by dilution in normal saline. It was found that the extract of the glandular lobe had little or no perceptible effect. The extract of the infundibular lobe had a distinct and remarkable effect on the heart-rate, which was slowed and augmented in force. At the same time the blood-pressure rises to a considerable extent, owing, apparently, to a peripheral constriction of the blood-vessels. The rise is followed by a temporary fall of pressure, during which the heart-rate may be increased. When the vagi are previously cut the slowing of the heart is much less marked.

AUTHOR'S EXPERIMENTS.

Results of the experiments made upon animals in the laboratory of Parke, Davis & Co. by Dr. E. M. Houghton, at the author's request. Three tracings were taken.

The extract was made so that one cubic centimeter represents one gramme of the pituitary body. Alcoholic and aqueous extracts were made; in the case of the alcoholic extract it was evaporated at low temperature and taken up with water.

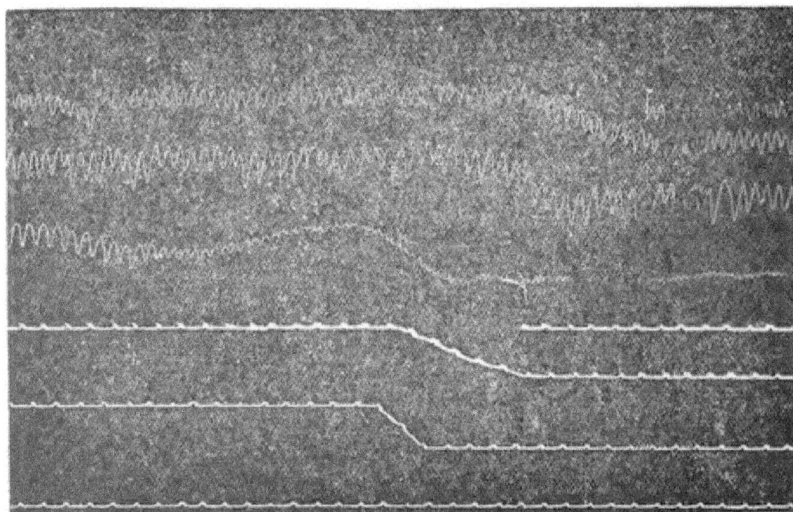

FIG. 26.—Effect of injection of filtered infusion of pituitary gland (½ grain). Star denotes moment of injection. Read from right to left. (Ott.)

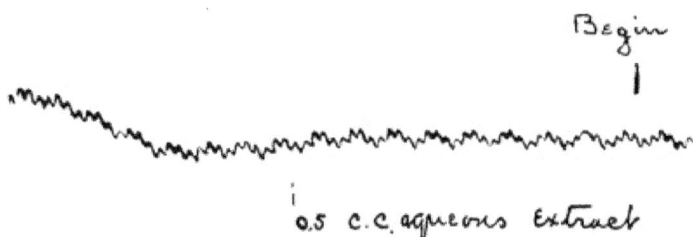

0.5 C.C. aqueous Extract

Fig. 27.—Action of the extract of pituitary gland on the circulation (Hinsdale).

Experiment 1.—Dog; weight, 9725 grammes. Injected intravenously one cubic centimeter of the aqueous extract of fresh pituitary gland. There was no immediate change in pressure, but at the end of one hour there was a gradual fall of ten millimeters.

Experiment 2.—Rabbit. Blood-pressure experiment. Aqueous extract of pituitary body injected intravenously.

Amount injected.	Blood-pressure.
	Normal, 83 mm. Hg.
.1 Cc.	Rose to 89 mm. Hg.; fell to 80 mm. Hg.
.5 Cc.	Rose to 89 mm. Hg.; remained at 87 mm. Hg.
.5 Cc.	Quick rise to 100 mm. Hg.; fell to 86 mm. Hg.
.25 Cc.	Gradual rise to 96 mm. Hg.; fell to 87 mm. Hg.
1. Cc.	Very gradual fall to 83 mm. Hg.
1. Cc.	Quick rise to 98, finally to 101 mm. Hg.

Experiment 3.—Rabbit. Blood-pressure experiment. Alcoholic extract of pituitary body injected intravenously.

Amount injected.	Blood-pressure.
	Normal, 90 mm. Hg.
1 Cc.	Immediate rise to 101 mm. Hg.; after ten minutes fell to 82 mm. Hg.
2 Cc.	Rose to 89 mm. Hg.; gradual fall to 79 mm. Hg.
2 Cc.	Gradual fall to 75 mm. Hg.
4 Cc.	Rise to 85 mm. Hg.
2 Cc.	Gradual fall to 77 mm. Hg.

Experiment 1 shows a slight fall in pressure.

Experiment 2 shows a primary rise, falling afterward to about normal.

Experiment 3 shows a primary rise, falling afterward 13 mm.

Treatment by Mercury and Iodides.—In a case in which there was no evidence of syphilis, antisyphilitic treatment was adopted by Schlesinger. There had been ptosis, complete paralysis of the oculo-motor nerve, and gray atrophy of the papilla with temporal hemianopsia and contraction of the preserved half of the visual field.

The ptosis disappeared. The paralysis of the third pair was relieved, and the visual field became normal, except for a scotoma for blue in the left eye. The papilla became nearly normal.

Operative Treatment.—This has been instituted in three cases, viz., those of Caton, Lynn Thomas, and Hare. In the first instance

Mr. Paul was asked to operate, and in February, 1893, he removed a portion of the right temporal bone to the extent of about three superficial inches. The brain was found in a state of great tension; it did not pulsate at all and bulged outward as much as the dura allowed. From the time of operation the intense neuralgic pain ceased, but the blindness, which was almost complete, soon became absolute. Death ensued three months later.

In the case of Thomas the skull was opened to the extent of three by four inches on the left side. The bone was thickened and softer than normal. The patient is said to have been transformed "from being a miserable imbecile into a useful, intelligent being." The headache was cured and power was restored to the bladder.

Hare's case was operated upon for the relief of intense neuralgia. Professor Keen resected the following nerves: The supra-orbital, supra-trochlear, auriculo-temporal, auricularis magnus, occipitalis major and minor. No improvement followed, however.

ACROMEGALY.

XX., aged 30; single; English; by occupation a bookkeeper.[1] Family history negative. Previous history: Excepting the usual diseases of childhood, the patient had always enjoyed good health until five years ago, when he is reported to have had pleurisy. He expectorated blood for one week. In the spring of 1891 he fell about twenty feet, sustaining a severe scalp wound in the occipital region and a contusion of the back. In 1893 patient was in good health, weighing about 175 pounds. He wore a No. 7 hat, a No. 7½ glove, and a No. 7½ shoe. He was not seen by his relatives until three years later, when he wore a No. 8½ hat, a No. 11 shoe, and had to have his gloves made to order. His weight at the time was about 250 pounds. Some time between 1893 and 1896 the patient developed acromegaly and was treated for it in Chicago. The course of his disease is said to have been arrested under treatment. About six months ago the patient was treated for syphilis. His symptoms disappeared shortly after medication commenced. Two months ago the patient's right eye began to trouble him, and he is said to have been treated for syphilitic iritis. One month ago the eye was still in bad condition. About this time his friends began to suspect him of using narcotics. He slept very little at night, but was drowsy at odd intervals during the day. Recently the patient has eaten very little on account of nausea and vomiting, but has been thirsty all the time. Thinking that his thirst was due to atropine that drug was omitted by his physician and duboisine used instead for his iritis.

[1] The notes of this case were kindly sent to the author by Dr. S. Baruch, of New York. Dr. Harlow Brooks, who made the autopsy, subsequently published an account in the *New York Medical Journal.*

Patient's mental condition has been irritable since May, 1896. He is a steady and heavy drinker.

Present condition: Fairly well nourished; tongue coated, brown, and dry; skin congested; extremities cold; appears to be in a stuporous condition, but answers questions; is suffering with dyspnea; heart enlarged, apex beat diffuse, no murmurs; lungs showed dulness, subcrepitant râles in left axillary line; pulse feeble; abdomen slightly distended; liver dulness normal; left pupil about normal, reacts to light; right pupil dilated and does not react to light; exophthalmus; haziness of right cornea and conjunctiva injected around cornea; has myopic astigmatism O.D. — .50 cyl. 1.50 ax. 15 deg. O.S. — .25 cyl. 1.35 ax. 165 deg.; peculiar, sweetish odor to breath; hands and fingers somewhat thickened; ends of fingers slightly clubbed; cheek bones prominent; end of nose enlarged; a small furuncle on right side over shoulder; temperature 99.2°; pulse 120; respiration 30; urine, straw, clear, acid, 1.023, albumen two per cent., sugar 7½ per cent.

Treatment.—Put to bed, milk diet. Given seven minims majendie at 11.20 A.M.

2.15 P.M. Catheterized and thirty-three fluidounces urine obtained, examined as above.

2.45 P.M. Is in a marked stupor. Given castor oil two ounces, with four ounces water.

4 P.M. Temperature 101°, pulse 132, respiration 26. Is restless and shouting out. Given liquid majendie 7½ minims.

6.30 P.M. Five ounces urine obtained by catheter.

8 P.M. Is in stuporous condition, completely unconscious, pulse very feeble. Given strychnine sulphate $\frac{1}{30}$ grain hypodermically. Temperature 103.8°, pulse 128, respiration 38.

8.15 P.M. Died (December 2, 1896).

Circumference of head at level of eyebrows		23½ inches.
Circumference of head at submento-bregmatic		27 inches.
Circumference of occipito-mental		26½ inches.
Breadth of base of nose		1⅜ inches.
Breadth across alæ		2 inches.
Circumference of neck		17 inches.
Circumference of right elbow		11 inches.

	Right.	Left.
Circumference of wrist	7⅞ inches.	7 inches.
Circumference of hand at knuckles	10½ inches.	10½ inches.
Circumference of base of thumb	3¼ inches.	3½ inches.
Circumference of last phalanx of thumb	3⅞ inches.	3¼ inches.
Length of middle finger from articulation of metacarpus	4½ inches.	4½ inches.
Circumference of middle finger	3½ inches.	3⅝ inches.
Length of forefinger	4 inches.	4½ inches.
Circumference of forefinger	3⅞ inches.	3¼ inches.
Length of ring finger	4½ inches.	4½ inches.

	Right.	Left.
Circumference of ring finger	3 inches.	3 inches.
Length of little finger	3¼ inches.	3½ inches.
Circumference of little finger	2¼ inches.	2¾ inches.
Circumference of groin	20¼ inches.	20½ inches.
Circumference of knee	15½ inches.	16 inches.
Circumference of calf	14 inches.	14½ inches.
Circumference of ankle	10¾ inches.	10⅝ inches.
Circumference across foot	10¼ inches.	10 inches.
Breadth across foot	4½ inches.	4½ inches.
Length of great toe	2¾ inches.	2⅝ inches.
Circumference of great toe	4½ inches.	4¼ inches.

General Description.—Head large, lips thick; eyebrows prominent and one hanging; eyelids thick and heavy; cheek bones prominent; space between the eye and malar bone one plane; nose short, thick, and flat, point of nose blunt and almost globular; tongue broad and thick; teeth good; ears large and prominent; neck short; shoulders broad and square; arm and forearm normal in appearance to wrist; bones at wrist-joint enlarged; hand short, broad, and thick; ends of fingers squared; nails broad, square, flat, and curved at ends, but not thickened; base of digits broad and flat; thighs and pelvis well formed; knee-joints prominent; lower epiphysis of tibia and fibula enlarged and prominent; foot drawn inward and upward, broad and flat; toes short, terminal phalanges broad and square, nails flat and not thickened.

General Musculature: Muscles large and soft, small amount of adipose tissue. Face had a peculiar edematous appearance. Chest well formed, abdomen distended and tympanitic. Left pleural cavity contained a few old adhesions; pericardium negative. Heart: small amount of ante-mortem clot in right ventricle; small amount of surrounding fat; muscle solid and firm, of a good color; normal; weight 15 ounces. Lungs slightly congested; bronchi congested; bronchial glands slightly enlarged, otherwise negative. Much gaseous distention of intestines. Acute peritonitis, slight degree, most marked about celiac axis, extending over toward left side. In the region slightly above sigmoid flexure was found an area of encysted exudate, not purulent. Spleen was adherent to cardiac end of stomach and elongated to a length of 6½ inches; weight 13 ounces. Capsule rough, covered by small fibrous masses, granular to the touch, light in color, flabby in consistency, not congested. Pancreas large, tissue very light in color. Stomach distended; large amount of acid-smelling, dark, undigested food; some large masses of cabbage (?). Stomach wall very normal in appearance, but dilated. Caput coli very much dilated with fecal matter; appendix long, adherent to posterior surface of cæcum; no intestinal lesions. Bladder normal in appearance, contained small amount of urine.

Liver large and light in color; arteries not thickened; substance friable and not granular; weight 8 pounds 1 ounce. Gall-bladder distended, duct patulous. Adrenals normal in appearance; right kidney fatty and granular; pelvis of left kidney dilated and distended with urine; pyramids of both kidneys conjested, markings distinct; capsules slightly adherent; vessel walls normal; pelvis dilated; cortex thin; united weight, 1½ pounds. Head: scar the size of a silver half dollar, one and one-half inches to the right of the median line, on a line extending from mastoid process to mastoid process; skull-cap greatly thickened, cancellous tissue small in amount; acute meningitis with effusion, most extensive over anterior lobes.

Pituitary Body: Rising from the pituitary fossa and inclining toward the left was found an ovoid red mass, measuring 1.5 centimeters antero-posteriorly and .7 centimeter vertically. It was attached below, apparently, to the pituitary body. It was of a soft and jelly-like consistence and quite vascular. It pressed directly on the left optic tract, just posterior to the chiasm. The tumor was attached to the hypophysis, which was enlarged to about five times its usual volume. The pituitary fossa was greatly enlarged and the bones comprising its wall were abnormally thin. No adhesions existed between the pituitary body or tumor.

SKELETON OF A GIANT.

Of the person whose skeleton I propose to describe, nothing is known, with the exception of one fact—that he was a native of Kentucky, U. S. A. In the year 1877, Prof. Joseph Leidy was informed by Prof. A. E. Foote that a body of a giant was offered for sale, provided no questions were asked which might lead to its identification. Arrangements were soon made by Dr. William Hunt through the gentlemen mentioned above, and the body was transferred to Philadelphia, where the skeleton was prepared and mounted by Mr. R. H. Nash.

All the persons mentioned are now dead. They were never able to ascertain, or, at least, thought it prudent not to make inquiry as to the antecedents of this giant, whose skeleton, the largest and most interesting in America, now adorns the Mütter Museum of the College of Physicians of Philadelphia.

It will be interesting to make a comparison of this skeleton, as a whole, and the individual bones composing it, with some of the famous giants of which we have records, and particularly with skeletons of giants in London and Dublin, and with the acromegalic skeleton in Edinburgh, which has attracted world-wide

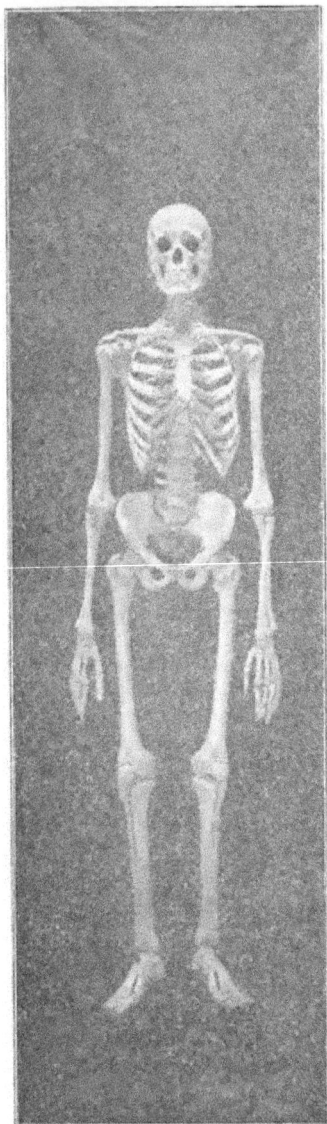

FIG. 28.—Normal skeleton. Height, 6 feet 1
inch. (From the private collection of Dr.
George McClellan.)

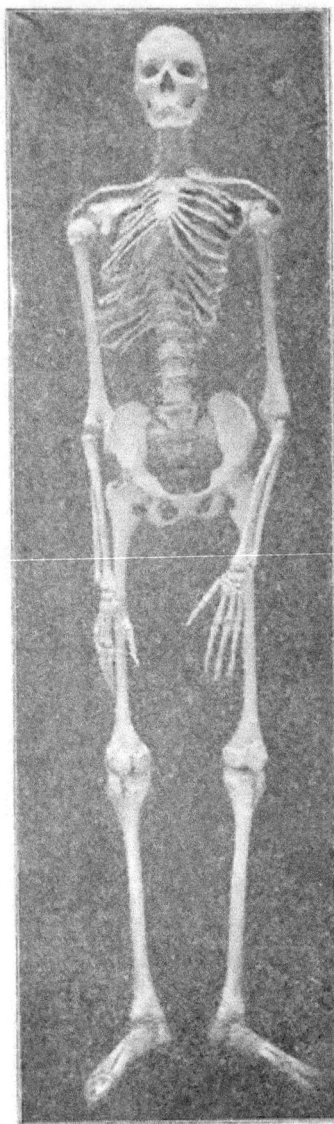

FIG. 29.—The American giant. Height, 7 feet
6 inches.

attention. The author is indebted to Professor Cunningham for measurements of these last named skeletons, and he has endeavored so to arrange his description as to conform to the model which Professor Cunningham has given. Not having been furnished with any name for the person whose skeleton I describe, I shall refer to him as the American Giant.

Height.—It is probable that the American Giant had attained the age of 22 or 24 at the time of his death. The bones seem to have attained their full development, although the epiphyseal junctions are plainly visible in all the long bones.

The spine has undergone a kyphoskoliosis which detracts considerably from the height which would otherwise exist. Taking the skeleton as we find it, the height is seven feet six inches. This measurement includes suitable artificial intervertebral disks, which were supplied in its preparation. There is a rule formulated by the late Prof. Sir George Humphrey[1] by which we may consider the height of any individual, under normal circumstances, to stand in proportion to the length of the femur as 1000 to 275. But in our case, as in other abnormal skeletons, this rule may lead us astray; whatever the error may be in short skeletons or those of moderate height, in the case of giants the disproportionate length of the long bones, particularly those of the lower extremity, used as a basis of calculation, will lead us to a slightly exaggerated estimate of the total height. Applying this rule, however, to the American Giant, we have R. 275: 1000: :650: *2363* — L. 275: 1000: :666: *2421*, which equals seven feet eight and three-quarter inches in case we estimate by the right femur, and seven feet and eleven inches in using the left femur.

The measurements, therefore, as given in the following table taken from the skeleton itself, doubtless underestimate the size of the giant in life, before kyphosis became extreme.

	Mm.	Ft.	In.
T. Au Austrian, de l'Expos. Sc. Anth. de Paris	2550	8	4¾
R. Marianne Wehde	2550	8	4¾
T. A Kalmuck, Orfila Museum	2530	8	3½
T. A Swedish guard of Frederick II	2520	8	3½
R. Chang	2360	7	8¾
C. Byrne, R.C.S.E.	2310	7	7
R. Drasal (Olmutz)	2300	7	6½
A. The American Giant	2295	7	6
V. Winkelmeier, born in Upper Austria	2278	7	5¼
R. Thos. Hasler, Bavaria (acromegalic?)	2270	7	5¼
A. Henry Alexander Cooper (Yorkshire Giant)	2226	7	5
L. Innsbruck Giant	2226	7	3½
V. Murphy (Irish), Marseilles Museum	2220	7	3½
L. Berlin (Cat. No. 3040), one of the famous guards of Frederick II	2220	7	2½
L. Lolly, Pomeranian, St. Petersburg	2195	7	2½
C. Magrath (Irish Giant), Dublin	2177	7	2¼

T., Topinard; C., Cunningham; L., Langer; R., Ranke; V., Virchow; A., Author.

[1] Human Skeleton, 1858, Table IV, page 108.

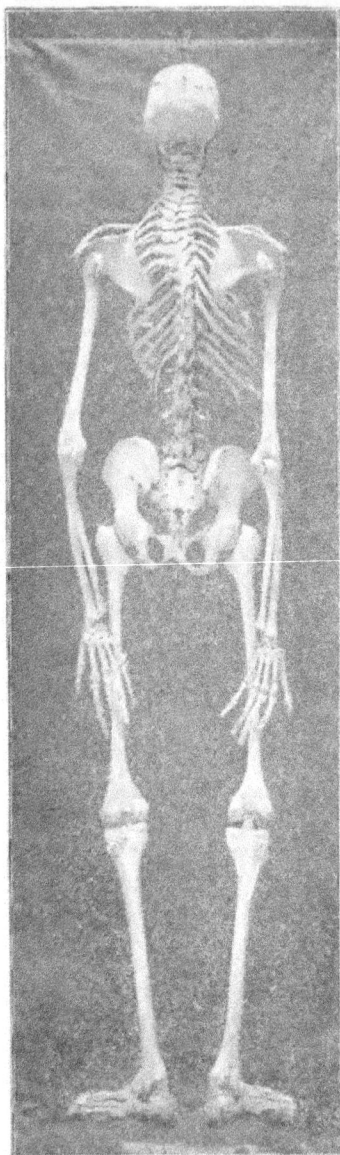

FIG. 30.—American giant; posterior aspect.

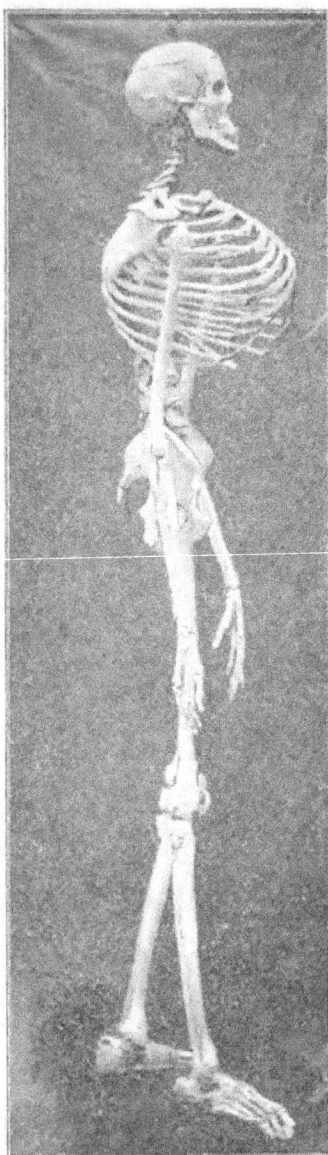

FIG. 31.—American giant; side view.

Bones.—The Skull: The skull of the American Giant bears a fair proportion to the great size of the skeleton. There has been some discussion among craniologists as to the usual ratio of the size of the skull to that of the skeleton in the case of giants. Virchow maintains that the horizontal circumference and the various diameters of the cranium exceed the average. He states, however, that the basis cranii is relatively short. Langer holds that the head in giants is, as a rule, relatively small.

Skull measurement.	American Giant.	Irish Giant.	Byrne, R. C. S. E.	Edinburgh acromegalic sk.
Cubic capacity, Cc	2320	1600	1520	1580
Length, glabella-occipital	234	198	215	200
internal	195	......	......	......
Height, basi-bregmatic	......	139	148	142
binaural over bregma	354	......	......	......
Vertical index	78.3	70.2	68.8	71
Breadth, maximum	145	155	151	148
minimum frontal	......	......	......	......
Cephalic index	......	78.5	70.2	74
Horizontal circumference	640	568	593	561
Length of foramen magnum	51	46	43	33
Breadth of foramen magnum	39	......	......	......
Interzygomatic breadth	147	156	148	150
Intermalar breadth	133	......	......	136
Facial index	......	61.5	55.4	54
Orbital width	52	44	42	45
Orbital height	42	43	34	36
Orbital index	80	97.7	81	80
Palato-maxillary length	......	50	61	67
Palato-maxillary breadth	......	65	62	70
Palato-maxillary index	......	130	101.6	104.4

The measurements, therefore, show that we have to do with a truly gigantic skull, its internal capacity being nearly one-half again as large as in the case of the skeletons in London, Dublin, and Edinburgh. The increased capacity is chiefly due to increased length, by reason of which the skull is classed as dolicho-cephalic. The interzygomatic diameter is slightly under that of the three skulls with which we have compared it. The intermalar breadth, on the other hand, is in excess, due largely to the great development of the antrum. The frontal sinuses have the following diameters: Transverse, 7.5; vertical, 7; antero-posterior, supra-orbital, 2.4.

Pituitary fossa.	American Giant.	Magrath.	Edinburgh skeleton.
Length	27	38	22.5
Depth	17	28	28
Breadth	42	...	21

Face.—The face is large, even in proportion to the large cranium. The air sinuses are large and give the great intermalar measurement, and the inferior maxilla, slightly prognathous, is massive.

	Length of face	Stature.	Face-stature index.
American Giant	148	2285	6.48
Irish Giant	156	2177	7.16
Winkelmeier	149	2278	6.54
Edinburgh acromegalic	148	1830	8.08
Murphy	143	2220	6.44
Byrne, R. C. S. E.	137	2287	5.99
Normal	120	1710	7.01

	Naso-alveolar length.	Naso-alveolar stature index.
American Giant	90	3.93
Irish Giant	96	4.40
Byrne, R. C. S. E.	82	3.58
Edinburgh acromegalic	81	4.42
Normal	73	4.26

The mandible is a large bone, as will be seen from the following measurements: Intercondyloid, 141; intergonial, 113; mento-alveolar, 40; width at angle, 33; width of ramus, 40; angle of ramus, 140°.

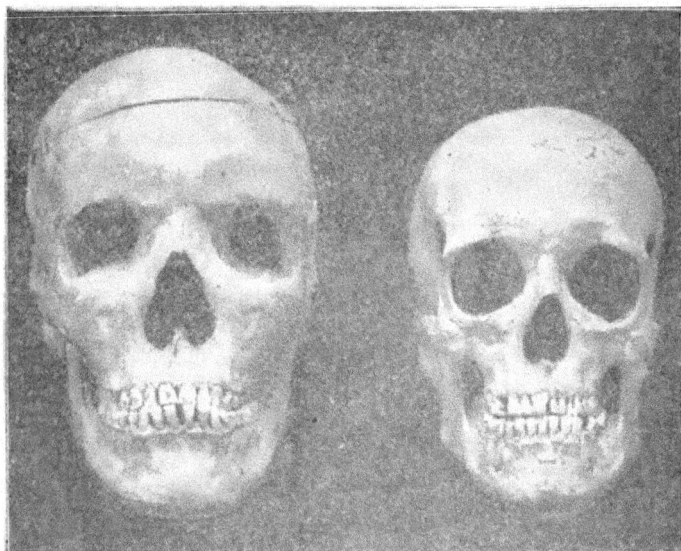

FIG. 32 —Skull of the American Giant compared with a normal skull.

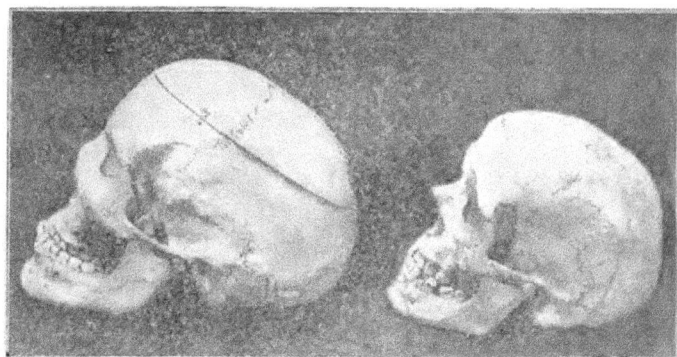

The coronoid process rises higher than the condyloid process. The lower jaw is prognathous, so that the four incisor teeth project slightly in advance of the corresponding teeth above. All the teeth in both jaws are in place and nearly all are incrusted with tartar.

	Length of the face from nasion to chin compared with the size of the cranium. Circumference, 100.	Naso-alveolar length compared with size of the cranium. Circumference, 100.
American Giant	39 23.1	14.0
Irish Giant	27.4	16.8
Edinburgh skull	26.3	14.4
Byrne, R. C. S. E.	23.1	13.8
Average	22.8	13.8

These indices show that there is a slight excess in both the maxillary and mandibular portions of the face. The alveolar portions of the jaw are perfect, having undergone no absorption.

Relation of maxillary to mandibular portions of the face; symphyseal height, 100: American Giant, 44; Irish Giant, 47.9; Byrne, R. S. C. E., 50; Edinburgh skull, 56.4.

The orbits are extremely capacious.

Vertebral Column.—The vertebræ are subject to conspicuous alteration, but they have been so mounted as to give a correct representation of the curves which existed in life. As usual in acromegaly, we find a kyphoskoliosis, in which, however, the cervical vertebræ do not partake.

Viewed antero-posteriorly, we find a sharp curve in the dorsal and lumbar region, with the convexity to the right. The kyphosis reaches its maximum at the ninth thoracic vertebra, resulting in a compression and absorption of the body of that vertebra. The following are the measurements of the anterior surface of the vertebral bodies: Second thoracic, 30 millimeters; third thoracic, 28 millimeters; seventh thoracic, 25 millimeters; ninth thoracic, 10 millimeters; eleventh thoracic, 31 millimeters; first lumbar, 36 millimeters; second lumbar, 43 millimeters; third lumbar, 45 millimeters; fourth lumbar, 47 millimeters; fifth lumbar, 50 millimeters. Greatest width of the first thoracic vertebra, 97 millimeters (3.75 inches). From the atlas to the promontory of the sacrum, 820 millimeters; from the same points in a direct line, 750 millimeters; from the atlas to the tip of the coccyx, along the anterior borders of the vertebræ, 1030 millimeters.

Viewed laterally, we find the first four thoracic vertebral spines prominent in the convexity; next, the right transverse processes of the sixth, seventh, eighth, ninth, tenth and eleventh vertebræ; then the spinous processes of the twelfth thoracic and the lumbar vertebræ. The great lateral displacement of the vertebræ brings the posterior margin of the right scapula seven centimeters behind the corresponding margin of the left, and in line with the right transverse process of the seventh thoracic vertebra.

The lumbar vertebræ are massive. The sacrum is composed of four vertebræ, instead of the usual number, five. Its width is 16 centimeters; length, 13.2.

The coccyx consists of three vertebræ, instead of the usual number, four.

The ribs are long and narrow, and relatively straight. The seventh and eighth right ribs, measured along the under border, are 45 centimeters and 43.5 centimeters in length; the sixth rib on the left side on the outer side is 43.5 centimeters long; the seventh, measured along the under side, is 43.8 centimeters in length.

The sternum has a total length of 23.5 centimeters. It is a large, well-proportioned bone.

	Length.	Width.
Manubrium	6.7 cm.	8.4
Gladiolus, 1st segment	4	4
remainder	8.9	5.8
Ensiform cartilage	2	....

The thorax is large, but narrow in proportion to its depth. Girth, 109.7 centimeters (43¼ inches); antero-posterior diameter, 43 centimeters; lateral diameter, 28 centimeters.

This corresponds with Langer's observation that in individuals of great stature the thorax is relatively narrow. In this respect the American Giant differs from the Irish Giant, Magrath, whose broad chest and boldly curved ribs give a girth of 132 centimeters (52⅛ inches).

Pelvis.—The pelvis is large and proportionate to the size of the skeleton. The bones are thickened at their borders and bear the marks of periosteal inflammation. Especially is this noticeable in the iliac crests and above the acetabula, which are evidently arthritic. The conformation is of the rachitic type, judging from the increase of the measurement between the anterior superior spines and the crests.

	Centimeters.
Between the anterior superior spines	35.5
Between the crests	33.5
Between the middle points of the ischii	15.5
Antero-posterior diameter of the pelvic inlet	14.
Right anterior oblique diameter of the pelvic inlet	17.5
Left anterior oblique diameter of the pelvic inlet	17.7
Depth of pelvic cavity	......
Height of pelvis	......

This pelvis is thus somewhat smaller than that of Magrath (Dublin) and of Byrne (R.C.S.E.). The latter have a breadth of 38.6 and 38 centimeters respectively.

The cavity of the pelvis is considerably encroached upon by the convexities which mark the position of the acetabula, particularly that of the left side. The acetabula are very deep and separated by thin bone from the pelvic cavity. The bodies of the iliac bones are exceedingly thin.

Upper Extremity.—The clavicles: Length, 212 millimeters. The scapulæ: Length, acromion to angle, R. 268, L. 268 millimeters; breadth, R. 144, L. 150.

The Humeri.—The right humerus is 47.5 centimeters long and has a circular perforation in the olecranon fossa 16 millimeters in diameter.

	Length of humerus.		Relation of humerus to stature. (Stature = 100.)	
	Right.	Left.	Right.	Left.
American Giant	47.5	......	......	......
Irish Giant	43.1	43.3	19.7	19.8
Byrne	45.0	45.0	19.0	18.1
Edinburgh skeleton	34.8	33.1	19.6	18.8

Average, 98.8, Topinard.

Ulna.—Right ulna 37.8, left ulna 37.5 centimeters long.

Radius.—Right, 354; left, 360.

Hand.—Length of hand from scaphoid to tip of middle finger is 25 centimeters.

	Length of hand.	Relation to stature. (Stature = 100.)
American Giant	25.0 centimeters.	10.9
Irish Giant	25.8 centimeters.	11.8
Byrne (R. C. S. E.)	26.3 centimeters.	11.3

Lower Extremity. — *Femora:* The femora, though slender, in proportion to the great size of the body, are symmetrical, and the shafts are well formed and not unduly curved. They vary 11 millimeters in length; the right is 65.5 centimeters in length; the left, 66.6. The shafts are 84 and 85 millimeters in circumference for the right and left sides respectively, and 20 millimeters in diameter. The shafts are thus no thicker than in ordinary individuals. The hip-joints are markedly arthritic. The femoral heads are misshapen and the necks are deformed. The latter, instead of being circular in shape on cross-section, are semilunar, with the flat side anterior; they are short and, instead of being placed obliquely, approach a right angle in relation to the shaft. The circumference and greatest diameter of the necks are as follows: Circumference, right, 13.5; left, 13. Greatest diameter, right, 5.7; left, 5.5.

The condyles are well formed and bear a proper relation to the tibiæ. The compact tissue is very thin at the extremities, barely covering the cancellous structure.

	Length of femur.		Girth of shaft.	
	Right.	Left.	Right.	Left.
American Giant	65.5	66.6	84 mm.	85 mm.
Irish Giant (Magrath)	60.3	62.4	......	......
Byrne, R.C.S.E.	62.5	64.2	......	......
Edinburgh skeleton	47.9	48.7	......	......

Tibiæ and Fibulæ.—The fibulæ curve with their convexity backward, so that the perpendicular to the chord of the arc is 47 millimeters. The middle portion of the shaft in both cases lies entirely posterior to the tibiæ.

	Length of tibia.		Relation of tibia to stature. (Stature = 100.)	
	Right.	Left.	Right.	Left.
American Giant	55.5	56.5	24.3	24.7
Irish Giant (Magrath)	50.6	50.4	23.2	23.1
Byrne, R.C.S.E.	54.1	53.7	23.6	23.4
Edinburgh skeleton	40.2	39.2	21.9	21.4

THE FOOT.

	Actual length.		Compared with skeleton. (Stature = 100.)
	cm.	in.	
American Giant	30.6	12	13.4
Irish Giant (Magrath)	30.0	11.75	13.7
Byrne, R.C.S.E.	31.7	12.50	13.8
Murphy[1]	31.0	12.25	13.9
Winkelmeier[1]	35.8	14.12	15.7
J. W. Walker's case[1]	36.0	14.13	18.3

[1] Measured during life.

BIBLIOGRAPHY.

Abram, John Hill: "A Note on the Development and the Anatomy of the Thyroid Gland in Disease," Pathological Reports, University Coll., Liverpool; *Liverpool Medico-Chirurgical Journal*, July, 1896.

Adami, J. George: "Internal Secretory Activity of Glands," *Montreal Medical Journal*, May, 1897.

Adler, I.: *Boston Medical and Surgical Journal*, 1888, p. 507, vol. cxix; *Medical News*, 1888, vol. lviii, p. 482; *New Yorker Medicinische Monatschrift*, May, 1889, p. 225.

Alibert: Précis théorique et practique des maladies de la peau, 1892, tome iii, 317.

Andriezen: "Morphology, Origin and Evolution of Function of the Pituitary Body and its Relation to the Central Nervous System," *British Medical Journal*, Jan. 13, 1894.

Appleyard: *Lancet*, 1892, i, 746.

Arnold: *Beiträge z. Path. Anat.*, 1891, x, p. 1; *Virchow's Archiv f. Path. Anat.*, vol. cxxxv, p. 1.

Ascher: *Neurologisches Centralbl.*, 13, 1894, p. 429.

Asmus, R.: *Arch. f. Ophth.*, xxxix, 1893, 3.

Bailey, Pearce: Case of acromegaly, with autopsy, *Philadelphia Medical Journal*, April 30, 1898.

Balzer: *Bull. et Mem. Soc. des Hôp. d. Paris*, 1892, ix, 237.

Barclay, John, and Symmes, Wm. S.: *British Medical Journal*, 1892, ii, 1227.

Dard: *Lyon Méd.*, 1892, lxix, 547.

Baruch, Simon: *Illustrated Medicine and Surgery*, New York, 1892.

Barrs: *Lancet*, 1892, i, 683.

Bassi: "Acromegaly with Syringomyelia and Cerebellar Tubercle; Autopsy," *Atti della Reale Accademia Lucchese di Scienze*, March 27, 1896.

Bayer, A.: *Tageblatt der Versammlung Deutscher Naturforscher*, etc., Wien, 1894, p. 309.

Becker: *Neurologisches Centralbl.*, 13, 1894, p. 503.

Beadles, Cecil F.: "Osteitis Deformans and Akromegaly," *Edinburgh Medical Journal*, March, 1898.

Benson, A.: *Dublin Journal of Medical Science*, 1895, c, 391–400; *British Medical Journal*, 1895, p. 949.

Berkley, H. J.: Case in a negro, *Johns Hopkins Hospital Bulletin*, September, 1891.

Bertrand, L. E.: *Revue de Méd.*, Paris, 1895, xv, 118.

Bettencourt, R.: *Jour. de Sc. Méd. de Lisbon*, 1890, liv, 366.

Bier, A.: *Mittheil a d. Chir. Klin. zu Kiel*, 1888, 196–203.

Bignami, A.: *Bull. d. Soc. Lancisiani di Osp. di Roma*, 1889, 1890, 1891, v, 238.

Bollinger, O.: *Aerztl. Verein., Münchener Med. Woch.*, 1893, p. 392.

Boltz, R.: *Jahresb. d. Schles. Gesellsch. f. Vaterl. Kult.*, 1891, Breslau 1892, lxix Abth., 95; *Deutsche Med. Woch.*, 1892, xviii, 635; *Jahresb. der Hamburg. Staatskrankenanst*, iii, 1894.

Bonardi, E.: "Un Caso, etc.; Autopsia," *Arch. Ital. di Chir. Med. Milano*, 1893, xxxii, 356; *Riforma Medica*, Naples, 1893, ii.

Booth, A.: Tr. New York Neurolog. Soc., *Journal of Nervous and Mental Disease*, 1893, p. 587.

Bourneville et Regnault: *Bull. Soc. Anatomique de Paris*, 1896, p. 587.

Boyce and Beadles: "Enlargement of the Pituitary Gland in Myxœdema," *Journal of Bacteriology and Pathology*, vol. i, pp. 223 and 359, 1892, and February, 1893; Path. Reports, University Coll., London, vol. i, 1892-93.

Bra, M.: La thérapeutique des tissus: Compendium des Médications par les extraits d'organes animaux, Paris, 1895, pp. 388-393.

Bradford: *British Medical Journal*, Dec. 26, 1891.

Bramwell, Byrom: Atlas of Clinical Medicine, vol. ii, part 3, 1892; Case in a giantess, *Edinburgh Medical Journal*, xxxix, p. 642.

Brigidi: *Società Medico-fisica Florentina*, Aug. 26, 1877.

Brissaud, E.: *Revue Neurologique*, 1893, 1-55.

Brissaud, E., and Meige, H.: "Deux cas de Gigantism suivi d'acromegalie," *Nouvelle Iconog.*, t. x, No. 6, 1897; "Gigantism et Acromegalie," *Rev. Scientifique*, Paris, 1895, 4th series, iii, 330; *Journal de Méd. et de Chirurgie Pratiques*, 1895.

Broadbent, W. H.: "Skiagraphs," *Lancet*, March 28, 1896, p. 846.

Broca, A.: "Un Squelette d'Acromegalie," *Arch. Gén. de Méd.*, Paris, 1888, il, 656.

Brooks, Harlow: "A Case of Acromegalia, with Autopsy," *New York Medical Journal*, March 27, 1897.

Brown, F. G.: *Trans. Hunterian Soc.*, London, 1892-3, 55; *British Medical Journal*, 1892, i, p. 862.

Brown, S.: *Chicago Clinical Review*, iii, July, 1894, 575; *North American Practitioner*, 1895, vii, 387.

Brown-Sequard, E.: Soc. de Biologie, Paris, '893, May 20.

Bruns: *Neurologisches Centralb.*, December, 1895, p. 1173; *British Medical Journal*, Jan. 18, 1896; Schilddrusentherapie, *Encyclop. Jahrbücher*, vi, Wien und Leipzig, 1896.

Bruzzi, A.: *Gazz. d. Osp.*, Mil., 1892, xiii, 866.

Bullard, E. L.: *Medical and Surgical Reporter*, 1895, lxxii, 591; *Journal Nervous and Mental Disease*, vol. xx, p. 743.

Bury, J. S.: *Medical Chronicle*, Manchester, July, 1891; *Brit. Med. Jour.*, 1891, i, p. 1179.

Buschan, G.: Myxœdema und Verwandte Zustände Zugleich ein Beitrag zur Schilddrüsen Physiologie und Schilddrüsen Therapie, Leipzig, 1896; 182 pp.; complete bibliography.

Buxer, C.: *Aerztl. Rundschau*, Munich, 1892, ii, 509.

Cabot, R. C.: "The Clinical Uses of the Preparations from the Thyroid Gland, Pituitary Body, etc.," Med. Communications to the Massachusetts Med. Soc., Boston, 1896, xvii, No. 1, 243.

Campbell, E. K.: Tr. Clinical Soc. of London, April 24, 1896; *British Medical Journal*, 1896, i, p. 1091; *British Medical Journal*, 1895, i, p. 81; Tr. Clinical Soc. of Lond., 1890, xxiii, p. 257, and 1894.

Carpenter, H. W.: *Journal of the American Medical Association*, 1897, i, p. 1043.

Carr-White: *Edinburgh Medical Journal*, 1889.

Caton, R.: "Treated by Operation," *British Medical Journal*, Dec. 30, 1893, p. 1421; *Liverpool Medico-Chirurgical Journal*, 1893, 369.

Caton, R., and Paul, F. T.: "Acromegaly Treated by Operation," *British Medical Journal*, 1893, ii, 1421.

Cattell, H. W.: "Skiagraph of an Acromegalic Hand," *International Clinics*, July, 1896.

Cenas: "Supposed Congenital Case," *La Loire Médicale*, Dec. 15, 1890.

Cepeda, G.: *Rev. Balear de Cienc. Méd.*, Palma de Mallorca, 1892, viii, 7.

Chadbourne, T. L.: "A Case of Acromegaly with Diabetes," *N. Y. Med. Jour.*, April 2, 1898.

Chalk: Trans. Path. Soc. of London, 1857, vol. viii, p. 305.

Chantemesse, M.: "Sur un cas de Syringomyelie à forme Acromegalique," *Le Progrès Méd.*, 1891, i, p. 273.

Chappell, W. F.: "Larynx in Acromegaly," *Amer. Med. Surg. Bulletin*, Jan. 18, 1896; Allbutt's System of Medicine, vol. v, p. 806.

Chauffard, A.: *Abeille Méd.*, 1895, lii, 244; *Bull. et Mém. Soc. Méd. de Hôp.*, 1895, 35, xii, 542; "Acromegalie fruste avec macroglossie," *La Semaine Médicale*, 1895, p. 305.

Chealdie et Lannois: De la cachexie pachydermique.

Chéron, P.: *L'Union Médicale*, Jan. 6, 8, 1891.

Church, A., and Hessert, W.: Medical Record, 1893, xliii, p. 545.

Clair-Symmers: British Medical Journal, 1891.

Claus, A.: Ann. Soc. de Méd. de Gand, 1890, vol. lxix; and 1893, vol. lxxii.

Claus, A., and Van der Stricht, O.: Ibid., 1893, p. 71.

Cohen, S. S.: Tr. Coll. Phys. Phila., 1892; Medical News, 1892, p. 518; Alienist and Neurologist, 1894, 374; College and Clinical Record, 1894, 112; International Clinics, 1894, ii, 57.

Collins, J.: Journal of Nervous and Mental Disease, 1893, xx, pp. 45-129, and p. 917.

Comini, E.: Archivio per de Sc. Med., vol. xx, Fasc. iv, 1896.

Costanzo, Feliciano: Rivista Veneta di Scienze Mediche, xxii, 1895, Venezia; Centralb. f. Nervenheilkunde, xviii, 1895, p. 459.

Crego, F. S.: Medical Record, 1894, vol. xlv, p. 215.

Crowell, G.: Trans. Ophth. Soc. United Kingdom, Lond., 1890-91, xi, 84-86.

Crétien: Rev. de Méd., 1893.

Cunningham, D. J.: The Irish Giant (Acromegalic), Trans. Royal Irish Academy, vol. xxix, part xvi, 1891.

Cunningham-Thomson: Journal of Anatomy and Physiology, 1889-90, vol. xxiv, p. 475.

Dallemagne: Arch. de Méd. Expér. et d'Anat. Path., 1895, vii, 589. Three cases with autopsy.

Dalton, N.: "Acromegaly and Myxœdema," Lancet, 1897, ii, 1190; Trans. Pathol. Soc., Lond., 1897.

Dana, C. L.: Journal of Nervous and Mental Disease, 1893, xx, 725, and 1894, p. 141.

Day, F. L.: Trans. Rhode Island Med. Soc., 1889-93, iv, 541; Boston Medical and Surgical Journal, 1893, p. 725.

Dehierre: Rev. Gén. d'Ophtal, Paris, 1891, x, 12.

Denti: Atti di Assoc. Med. Lomb. Milano, 1892, 41; Annali d'Ottalm., xxv, 1896, Fasc. 6.

Dercum, F. X.: American Journal of the Medical Sciences, Philadelphia, 1893, cv, 268.

D'Esterre, John Norcott: "Case of Acromegaly," British Medical Journal, Dec. 4, 1897.

Dethlefsen: Med. Aarsskr. Kjøben, 1892, 83-88.

Dinke: (See Vinke).

Dodgson, R. W.: Lancet, 1896, i, 772; British Medical Journal, March 14, 1896.

Daebbelin, C.: "Pseudo-acromegalie und Acromegalie," Thesis Königsberg, 1895.

Doyne: "Optic Atrophy in Acromegaly, with Charts of Fields of Vision," Trans. Ophthal. Soc. United Kingdom, 1895.

Dreschfeld, J.: British Medical Journal, Jan. 6, 1894, p. 4.

Du Cazal: Bull. et Mém. Soc. Méd. d. Hôp. de Paris, 1891, viii, 435; Progrès Médical, 1891, ii, pp. 295 and 585; La Semaine Médicale, Paris, Oct. 21, 1891.

Duchesneau, G.: Paris, 1892, De L'Acromégalie.

Dulles, C. W.: Medical News, 1892, No. 5.

Dyson, W.: Quarterly Medical Journal, January, 1894.

Editorial, Medical News, April 9, 1892.

Ellinwood: San Francisco Lancet, 1883, p. 159.

Elliott: Lancet, 1888.

Erb: "Ueber Krankhaften Riesenwuchs," Deutsches Archiv f. Klin. Med., 1888, Bd. 42; Naturforscherversammlung in Heidelberg, September, 1889, No. 62; Münchener Med. Woch., 1894, No. 24.

Eschle: Therap. Monats., January, 1896.

Eshner, A. A.: Medical News, 1895, lxvii, 458.

Ewald: Berl. Klin. Woch., March 18, 1888; Virchow's Archiv, Bd. lvi, 1892.

Farge: Prog. Méd., 1889, 2d series, x, July 6.

Fazio, F.: La Riforma Med., 1896, ii, p. 399.

Feld, F. A.: Case showing acromegaly and hypertrophic pulmonary arthropathy. British Medical Journal, July 1, 1893, p. 14.

Finlayson, James: International Clinics, October, 1896.

Flemming, Percy: Trans. Clinical Soc. of London, 1890.

Fournier, J. B. C.: "Acromegalie et troubles cardio-vasculaires," Thèse de Paris, 1896.

Foy, G.: "Cheiromegaly," Medical Press and Circular, 1891, lii, 491.

Fraenkel, A.: Münchener Med. Woch., 1897, p. 401.

Fraeutzel, O.: Deutsche Med. Woch., 1888, xiv, 651.

Franke: Case with temporal hemianopsia, Klin. Monatsbl. f. Augenh., 1896, p. 259.

Fratnich, E.: Allgemeine Wiener Med. Zeitung, 1893, No. 40, and 1892, No. 37; Riv. Veneta di Sc. Med., Venezia, 1892, xvii, 238.

Freund, W. A.: Samml. Klin. Vort., 1889, Nos. 329 and 330; Rev. de Sciences Médicales, xxxiv, p. 569.

Friedreich: Virchow's Archiv, Bd. xlvii, 1868, 83. Hyperostosis, etc.

Fritsche und Klebs: *Ein Beitrag zur Pathologie des Riesenwuchses*, Leipzig, 1884.

Furnivall, Percy: *Lancet*, ii, 1897, p. 1190.

Gajkiewicz, W.: *Gaz. lek. Warsawa*, 1891, Nos. 43 and 44, 25, xi, and 1896, No. 37; "Pseudo-akromegalie und Akromegalie," Inaug. Dissert., Königsberg, 1895.

Galvani: *Rev. d'Orthop.*, Paris, 1895, vi.

Garnud et Arene: *Loire Méd. St. Etienne*, 1895, xiv.

Gaston et Brouardel: "Röntgen Photography in Acromegaly," *La Presse Méd.*, 1896, No. 61.

Gause: *Deutsche Med. Woch.*, 1892, xviii, 892.

Gauthier, G.: *Prog. Méd.*, 1890, 2 S., xi, 409-414, and 1892, No. 1.

Gerhardt: *Berliner Klin. Woch.*, Dec. 15, 1890.

Gessler, H.: *Med. Corresp.-Bl. d. Würtemb. arzt.*, Stuttgart, 1893, lxiii, 121.

Gley, E.: "Recherches sur la fonction de la glande thyroide," *Bull. de la Soc. de Biol.*, April 18, 1891; *Arch. de Physiol. Norm. et Path.*, 1892, pp. 81, 311, 435.

Godlee, R. J.: Trans. Clin. Soc. Lond., 1888, and *British Medical Journal*, 1888.

Goldsmith, C. P.: *Lancet*, 1896, i, 993.

Gonzalez Cepeda, J.: *Rev. balear de Cien. Med.*, Palma de Mallorca, 1892, viii.

Gordon-Brown: *British Medical Journal*, April 23, 1892.

Gardinier, H. C.: *Medical News*, 1895, lxvii, 262.

Gouraud, X.: *Bull. de la Soc. Méd. des Hôp.*, 1889.

Gubian: *Bull. d. Dispensaire de Lyon*, 1811.

Gorjatscheff: *Chir. Lepopisj.*, Bd. 1, 1892. Moskau.

Graham, J. E.: Trans. Ass. Amer. Phys., 1890, v; *Medical News*, Oct. 18, 1890.

Grasset et Rauzier: *Maladies du Systeme Nerveux*, 4th Ed., 1894, ii.

Griffith, Alex. Hill: Case of sarcoma of hypophysis, *British Medical Journal*, 1895, ii, 950.

Grocco: *Rev. gen. Ital.*, Pisa, 1891.

Grön, Kr.: "Hypertrophy of the Hypophysis in a Case of Myxœdema," *Norsk Magazin for Lægevidenskaben*, 1894, p. 734.

Gubian: *Bull. de Dispensaire de Lyon*, 1891, No. 16.

Guinon and Surmont: *Nouvelle Iconogr. de la Salpêtrière*, tome iii, Paris, 1890, pp. 147-160.

Guinon, G.: *Gaz. d. Hôp.*, Paris, 1889, lxii, 1161.

Gulliver: Tr. Path. Soc. Lond., vol. xxxvii, p. 511.

Hadden and Ballance: Trans. Clin. Soc. Lond., 1888, xxi, and 1885, xviii.

Haller: "Morphology of the Hypophysis Cerebri," *Morphol. Jahrbuch*, xxv, 1896.

Hanseman, D.: "Über Akromegalie," *Deutsche Medizinal-Zeitung*, Feb. 4, 1897; *Berliner Klin. Woch.*, 1897, No. 20.

Hare, H. A.: *Journal of Nervous and Mental Disease*, 1892, 250; *Medical News*, 1892, 257.

Harris, H. F.: *Medical News*, lxi, 520.

Hascovec, L.: *Rev. de Méd.*, March 10, 1893, p. 237; *Wiener Klin. Rund.*, April 28, 1895.

Higier: *Tageblatt der Versammlungen Deutscher Naturforscher*, etc., Wien, 1894, p. 309.

Hinsdale, Guy: MEDICINE, June, July, August, September, and October, 1898.

Henrot, H.: *Notes de Clinique Méd.*, Rheims, 1882 and 1887.

Hitschmann, R.: "Acromegalie und Augenbefunde," *Wiener Med. Club*, June 16, 1897; *Wiener Klin. Woch.*, 1897, p. 659.

Hertel, E.: *Arch. für Ophthal.*, 1895, xli, 1.

Herzog, B.: *Deutsche Med. Woch.*, 1894, 316.

Hofmeister: "Experimental Research on Ablation of the Thyroid," *Fortschr. d. Med.*, 1892, p. 81.

Holschewnikoff: "Acromegaly and Syringomyelia," *Virchow's Archiv*, vol. cxix, p. 10.

Holsti: *Festskrud fran Pathologisk-Anatomisk Institut.*, Helsingfors, 1890; *Zeitschr. für Klin. Med.*, Bd. xx.

Hornstein: *Virchow's Archiv*, Bd. xxxiii, 1893.

Horsley, V.: "On the Functions of the Thyroid," *British Medical Journal*, 1890, ii, 201, and 1892, i, 431.

Howell, W. H.: *Journal of Experimental Medicine*, vol. iii, No. 2, 1898.

Huchard, H.: "Anat. Path., lesions et troubles cardio-vasculaire de l'Acromegalie," *Journal des Praticiens*, 1895, ii, 249.

Hutchings, R. H.: "Acromegaly and Mental Disease," *Medical Standard*, Chicago, 1895, xvii, 129.

Hutchinson, Woods: "Acromegaly in a Giantess," *American Journal of the Medical Sciences*, 1895, cx; "Function of the Pituitary Body," *Medical News*, 1896, ii, p. 707; "The Pituitary Gland as a Factor in Acromegaly and Giantism," *New York Medical Journal*, March 12, April 2, 1898, *et seq.*

Hutchinson, J.: Three cases, *Archives of Surgery*, London, April, 1891, and 1889-90, i, pp. 141-148.

Jores: (Bonn.)

Jorge, R.: *Arch. de psichiat.*, Turin, 1894, xx, 412.

Kalindero: *La Roumanie Médicale*, 1894, p. 65.

Kanthack: *British Medical Journal*, July 25, 1891.

Karg: *Archiv f. Klin. Chirurg.*, Bd. xli, p. 101.

Keen, W. W.: "Excision of Nerves in Acromegaly," *International Clinics*, Philadelphia, 1893, 3 S., ii, 191.

Kerr: *Lancet*, 1893, ii, 1256.

Koscheeff, J.: *Protok. Zasaid. Obsh. Vrach. Viatke*, 1892, 13–16.

Kerner: *Vereinsblatt der Pfälzischen Aerzte*, Frankenthal, August, 1890.

Kinnicutt, F. P.: "Therapeutics of the Internal Secretions," *American Journal of the Medical Sciences*, July, 1897.

Klebs: *Allgem. Path.*, 1889, ii.

Klebs und Fritsche: (Gigantism), *Ein Beitrag zur Pathologie des Riesenwuchses*, Leipzig, Vogel, 1884.

Kojevnikoff, A. G.: *Med. Obozr. Moscan.*, 1893.

Lancereaux: "Acromegalic Trophoneuroses of the Extremities; its Coexistence with Exophthalmic Goitre and Glycosuria," *Medical Week*, Paris, 1895, iii, 109; *Anatomie Pathologique*, t. iii, 1 re Partie, 29.

Lathuraz: "Result of the Autopsy in Pechadre's Case," *Lyon Médicale*, 1893, vol. lxxiii, p. 443.

Lavielle, C.: *Mém. et. Bull. Soc. de Méd. et Chir. de Bordeaux*, 1895, xxiv, 1.

Levi: *Archives Gén. de Méd.*, 1896, ii, p. 579.

Lichtheim: *Deutsche Med. Woch.*, 1893.

Linstnayer, L: *Wiener Klin. Woch.*, 1894, vii, 294.

Litthauer, Max: *Deutsche Med. Woch.*, 1891, xvii, 1282.

Little: *British Medical Journal*, 1895, ii, p. 950.

Lombroso: "Caso Singolare de Macrosomia," *Virchow's Archiv*, Bd. xlvi, 1869, 253, and Bd. lvi, 1872; also originally, *Giornale Ital. delle Malattie Ven.*, 1868.

Long: *Lehigh Valley Medical Magazine*, April, 1891, U. S. A.

Lovegrace: *Lancet*, 1892, i, p. 91.

Luzet, C.: *Arch. Gén. de Méd.*, 1891, i, 194.

Machado, V.: "Semeiologia Radiographique de l'Acromegalia," *Revista Portugueza de Med. e Cirurg.*, t. iii, No. 30, Jan. 15, 1898.

Mackie-Whyte: *Lancet*, 1893, i, 12.

Magnus-Levy: *Münchener Med. Woch.*, 1897, p. 400.

Macroq: *British Medical Journal*, 1892.

Mairet and Bosc: *Arch. de Phys. Norm. et Path.*, 1896, vol. viii, p. 600.

Marie, P., and Souza-Leite: "Essays on Acromegaly; with Bibliography," London, 1891.

Marie et Marinesco: *Archives de Méd. Expériment et d'Anat. Path.*, 1891, p. 539.

Marie, P.: *Rev. de Méd.*, Paris, 1886, vi, 297–333; *N. Iconogr. de la Salpêtrière*, 1888, i, 173, and 1889, ii, 45, 96, 139, 188, 224, 327; *Progrès Méd.*, 1889, 2 s., ix, iii, 1579; *Lecons de Clinique Médicale*, Paris, 1896, p. 51; *Bull. Méd.*, 1889, iii, 1579; *Brain*, Lond., 1889, xii, 59; *Bull. et Mém. de la Soc. des Hôp.*, May 1, 1896 (Hands in Acrom.).

Marina: *Riforma Med.*, 1893, ix, "Hypertrophic Osteo-arthopathy and Acromegaly."

Marinesco, G.: "Röntgen Rays in Acromegaly," *Comptes rendus Hebdom. des Séances Société de Biologie*, June, 19, 1896, p. 615; *ibid.*, 1892, p. 599; *Arch. de Méd. Expér. et d'Anat. Path.*, 1891, iii, 539; "Trois Cas d'acromegalie traités par des tablettes de corps pituitaire," *Bull. et Mém. Soc. Méd. des Hôp.*, 1895, 35, xii, 715; *Verhandl. d. Internat. Med. Congr.*, 1890, Berlin, 1891, iv; *Semaine Méd.*, Nov. 13, 1895.

Marino: *Berliner Klin. Woch.*, 1894, p. 988.

Marzocchi et Antonini: "Un cas d'acromegalie particille," *Riforma Med.*, January 22, 1897, No. 17.

Masoin, E: "Aperçus générales sur la physiologie du corps thyroïde," *Revue de questions Scientif.*, April, 1894.

Massalongo, R.: *Rev. Neurologique*, Paris, 1895; *Riforma Med.*, Napoli, 1892, viii, p. 10, iii, 74, 87; "Hyperfunction der Hypophyse, Reiseuwuchs und Acromegalie," *Centralbl. f. Nervenheilk.*, June, 1895.

Maus, J.: "Glandula thyreoidea und Hypophysis cerebri mit Hinweis auf die mit demselben in Beziehung stebenden Krankheitserscheinungen," Dissert., Griefswald, 1895.

Matignon, J. J.: "Un cas d'acromegalo-gigantisme chez un Chinois," *La Médecine Moderne*, 1897, No. 89.

Medical News, 1892, ix, 237.

Mendel, E.: *Berl. Klin. Woch.*, 1895, No. 48, 1129.

Mevel, P.: "Contribution a l'étude des troubles Oculaire dans Acromegalie," Paris, 1894. Thesis.

Meyer, F.: Ein Fall von Akromegalie, Hamburg, 1894. Kieler Dissert.

Meyer (Paris): "Ocular Symptoms in Three Cases of Acromegaly," *British Medical Journal*, 1895, ii, 949.

Michel, Middleton: "Pathology of the Pituitary Body," *Charleston Medical Journal and Review*, S. C., vol. xv, March, 1860.

Middleton, G. S.: "A Marked Case of Acromegaly with Joint Affections," *Glasgow Medical Journal*, 1894, xli, 401; *ibid.*, 1895, xliv, 127.

Minkowski, O: *Berliner Klin. Woch.*, 1887, xxiv, 371, and xxi.

Mistre, A.: *Rev. de Cien. Med. Habana*, 1895, x, 209–216.

Moebius: *Schmidt's Jahrbücher*, 1892, p. 22.

Moncorvo: "Microcephalic Case," *Rev. Mens. d. Mal. de l'Enfau.*, Paris, 1892, x, 549.

Monteverdi et Torrachi: "Un caso di acromegalia con emianopsia bitemporale," *Rivista Sperimentale*, 1897, Reggio.

Morax: *Neurolog. Central Organ.*, 1892.

Moritz: *Tageblatt der Versammlung Deutscher Naturforscher*, Wien, 1894, p. 309.

Mosler, C. F.: "Pachyacrie," *Festschrift R. Virchow*, Berl., 1891, ii, 101: *Wien. Med. Bl.*, 1892, xv, 56, 70, 87, 105, 121, 136; *Schmidt's Jahrbücher*, 1893, No. 12, p. 237.

Mossé, A.: *Comptes Rend. Soc. de Biol.*, Paris, 1895, ii, 686; *La Semaine Médicale*, 1895, p. 468; *Mercredi Méd.*, Sept. 11, 1895.

Motais: *Prog. Méd.*, 1891, May 7; *Annales d'Oculistique*, Jan.-Feb., 1886, p. 47.

Moyer, Harold N.: *International Medical Magazine*, 1894-5, iii, 34.

Murray, F. W.: *Annals of Surgery*, Philadelphia, 1893, xvii, 700.

Murray, G. R.: "Clinical Remarks on Cases of Acromegaly and Osteo-arthropathy," *British Medical Journal*, 1895, i, 293; "Acromegaly with Exophthalmic Goitre," *Edinburgh Medical Journal*, February, 1897.

O'Connor, J. T.: *N. Amer. Jour. Homeop.*, N. Y., 1888, iii, 345.

Naunyn: *Vereinsbeilage der Deutschen Med. Woch.*, 1896, p. 87.

Nonne: *Vereinsbeilage der Deutschen Med. Woch.*, 1896, p. 14.

Ogata, K.: *Chugai Iji Shimpo*, Tokio, 1895, 354.

Olechnowicz, W.: *Gaz. lek. Warszawa*, 1894, xiv, 113.

Orsi, P.: *Gazz. Med. Lomb.*, Milano, 1892, li, 201.

Orbitlard: *Rev. de Méd.*, 1892.

Orchiuzzi, L.: *Incurabili*, Napoli, 1892, vii, 350, 522, 549.

Osler, W.: Principles and Practice of Medicine, 1895, p. 1047.

Osborne, O. T.: *American Journal of Medical Sciences*, 1892, p. 617; *Yale Medical Journal*, December, 1897; Trans. Association Amer. Phys., 1897; Buck's Reference Handbook of the Medical Sciences, Supplement, 1893.

Ott, Isaac: "Note on Animal Extracts," *Medical Bulletin*, Philadelphia, 1896, p. 371.

Packard, F. A.: Trans. Coll. Phys. Phila., 1892; *American Journal of the Medical Sciences*, 1892, ciii, 657.

Paget: *Lancet*, Jan. 31, 1891.

Panas: *British Medical Journal*, 1895, ii, p. 950.

Parinaud: *Soc. Francaise d'Ophtalmologie*, Seance du 2 Feurier, 1891.

Park, R.: *International Medical Magazine*, Philadelphia, 1895-6, iv, 431.

Parsons, R. L.: *Journal of Nervous and Mental Disease*, 1894, xxi, 717; *New York Medical Journal*, 1894, lix, 88.

Pechadre: *Rev. de Méd.*, 1890, x, 175; *Lyon Méd.*, 1893, vol. lxxiii, p. 443.

Pel: *Berliner Klin. Woch.*, No. 3, 1891.

Pershing, H. T.: *Journal of Nervous and Mental Disease*, 1894, xxi, 693; *International Medical Magazine*, 1894-5, iii, 327.

Peterson, Frederick: "Acromegaly Combined with Syringomyelia," *Medical Record*, 1893, xliv, 391.

Pflüger: *Rev. Gén. d'Ophthal.*, 1892, 7.

Phillips, S.: Trans. Med. Soc. Lond., 1891-2, xv, 455; *British Medical Journal*, Feb. 27, 1892.

Pick, A.: *Prag. Med. Woch.*, 1890, xv, 521.

Pinel-Maissonneuve, L.: *Bull. et Mém. Soc. Méd. des Hôp.*, 1891, viii, 137; *Bull. et Mém. Soc. Franc. d'Ophthal.*, 1891, ix, 310; *Archives d'Ophthalmologie*, July-August, 1891, p. 309.

Pineles, F.: "Akromegalie und Diabetes Mellitus," *Jahrbuch der Wiener k. k. Krank.*, 1897, Bd. iv.

Putnam, J. J.: "Cases of Myxœdema and Acromegalia Treated with Benefit with Sheep's Thyroid; Cachexias of the Thyroid," *American Journal of the Medical Sciences*, 1893, cvi, 125; *Boston Medical and Surgical Journal*, Feb. 15, 1894.

Querenghi et Beduschi: "Contrib. alla casuistica dell' acromegalia," *Annali de Ottamologia*, xxvi, Pavia.

Rake, Bevan: *British Medical Journal*, 1893, i, 518.

Rampoldi, V.: *Gazz. Med. Lomb.*, Milano, 1894, liii, 101.

Ransom, W. B.: *British Medical Journal*, 1895, i, 1259.

Rath: Ein Beitrag zur Casuistik der Hypophysistumoren, Göttingen, 1888, *Graefe's Archiv f. Oph.*, xxxiv, 1888.

Rath: Ein Beitrag zur Casuistik der Hypophysistumoren, Göttingen, 1888.

Rauzier, G.: *Montpelier Med. Supp.*, 1893, ii, 623; "Diagnosis from osteo-arthropathy pneumique," *Rev. de Méd.*, ii, 1891, p. 56.

Recklinghausen, Von: *Virchow's Arch. f. Path. Anat.*, 1890, cxix, 36.

Redmond: *Dublin Journal of Medical Science*, January, 1891.

Regnault: *Revue Neurologique*, Feb. 28, 1898.

Reimar, M.: "Ein Fall von Amenorrhea bei Acromegalie," Halle Dissert., 1893.

Remington, F.: Tr. Med. Soc. New York, 1894, 266.

Renner: *Vereinsbl. d. Pfälz. Aerzte*, Frankenthal, 1890, vi, 164.

Rieder: *Münchener Med. Woch.*, 1893, p. 391.

Riegel: *Deutsche Med. Woch.*, 1893, p. 776.

Rogowitsch, N.: "Die Veranderungen der Hypophyse nach Entfernung der Schildruss," *Beitr. z. Path. Anat. v. Zeigler*, 1889, Bd. iv; "Zur Physiologie der Schildruse," *Centralb. f. d. Med. Woch.*, 1886, p. 530; "Sur les effets de l'ablation du corps thyroide chez les animaux," *Arch. de Physiologie*, 1888, Bd. ii, p. 419.

Rolleston, H. D.: *British Medical Journal*, 1890; *Lancet*, April 25, 1896, and 1897, ii, p. 1190; "Treatment of Acromegaly by Extracts of Thyroid and Pituitary Glands," *Lancet*, Dec. 4, 1897, p. 1443.

Ross: *International Clinics*, vol. i, 1891.

Ross and Bury: (Post-mortem notes), *Lancet*, 1891, i, 1383.

Roth, V. K.: *Med. Obozr. Mosk.*, 1892, xxxviii, 561.

Roth: *Virchow's Archiv*, 1889.

Rothmell, J. R.: *Journal of the American Medical Association*, March 20, 1897. (Autopsy.)

Ruttle: *Medical Press and Circular*, London, 1891; *British Medical Journal*, March 28, 1891, p. 697.

Sacchi, E.: *Riv. Veneta di Sc. Med.*, Venezia, 1889, xi, 417.

Sacchi and Vassali: *Centralbl. f. Allgem. Path. Anat.*, May, 1894.

Salbey, M.: "Ein Fall von Sogenannter Akromegalie mit Diabetes Mellitus," Erlangen, 1889, Thesis, *Münch. Med. Woch.*, 1889.

Sarbo, A.: *Orvosi hcti*, Budapest, 1892, 136, 149; *Pest. Med.-Chir. Presse*, Budapest, 1892, xxxviii, 575.

Saucerotte: *Mélanges de Chirurgie*, 1801, Part 1, p. 407; referred to in *Revue de Médecine*, tome vi, 1886, p. 316.

Saundby, R.: *Illustrated Medical News*, London, 1889, ii, 195.

Schaefer and Oliver: *Journal of Physiology*, 1895, p. 277.

Schaposchnikoff, B. M.: *Med. Obozr. Mosk.*, 1889, xxxii, 865.

Schiff, A.: "Ablation of the Thyroid," *Rev. Méd. de la Suisse Romande*, Feb. and Aug., 1884; "Influence of Thyroid and Pituitary Glands on Metabolism," *Wiener Klin. Woch.*, March 25, 1897.

Schlesinger, H.: "Treatment by Mercury," *La Semaine Médicale*, 1895, p. 51; *Wiener Med. Presse*, 1895, p. 186; Two Cases, *Neurologisches Centralbl.*, xiii, 1894, p. 741; "Partielle Acromegalie," *Wiener Klin. Woch.*, 1897, p. 445; *Neurologisches Centralbl.*, 1894, p. 741.

Schmidt: "Skiagraphs of Acromegaly," MEDICINE, p. 549, 1897.

Schultze, F.: *Deutsche Med. Woch.*, 1889, xv, 981, and 1896, p. 407; *Deutsche Zeitschr. f. Nervenheilk.*, 11 Band, 1 and 2 Heft, 1897; Fourteenth Congress, etc., Wiesbaden, 1896.

Schwartz: *St. Petersb. Med. Woch.*, 1890.

Schwoner, J.: "Ueber hereditäre Akromegalie," *Zeit. f. Klin. Med.*, xxxii, Supp., p. 202.

Sears, G. G.: *Lancet*, 1896, vol. ii, p. 614; *Boston Medical and Surgical Journal*, July 2, 1896.

Shiach, S. A.: *Lancet*, 1893, ii, 369.

Sigurini et Capariacco: *Riforma Medica*, 1895, xi, 207.

Silcock and Campbell: Two Cases, Trans. Clinical Society of London, vol. xxiii, 1890.

Silva: "Caso di acromegalia con atrophia dei testicoli," *La Riforma Med.*, 1895, ii, 532.

Snell: *British Medical Journal*, 1895, ii, p. 959.

Sollier: *France Méd.*, 1889.

Somers: *Occidental Medical Times*, Sacramento, California, October, 1891.

Souques: "Maccus, polichinelle et l'acromegalie," *Nouvelle Iconographie de la Salpêtrière*, No. 6, 1896.

Souques et Gasne: *Nouv. Iconog. de la Salpêtrière*, 1892.

Souza-Leite, J. D.: Essay, Paris, 1890.

Spillmann, P., and Haushalter: *Rev. de Méd.*, Paris, 1891, xi, 775.

Squance, T. C.: "Notes on a Post-mortem Examination," *British Medical Journal*, 1893, ii, 993.

Steinhaus: *Mém. de la Soc. de Méd. de Varsovie*, 1895, iv, p. 953.

Stembo: *St. Petersburg Med. Woch.*, 1891, Nos. 45 and 46.

Sternberg, M.: *Zeitschr. für Klin. Med.*, Berlin, 1895, xxvii, 86, 150; *Nothnagel's Specielle Pathologie*, Bd. vii, p. 116, Vienna, 1897; *Neurologisches Centralbl.*, xiii, 1894, p. 742.

Stieda: "Ueber das Verhalten der Hypophyse des Kaninchens nach entfernung der Schildruse," Dissert., Königsberg, 1890; *Ziegler's Beiträge z. Path. Anat.*, 1890, Bd. vii, p. 537.

Stroebe: *Centralbl. f. Pathologie*, vi, 1895, p. 721.

Strümpell: *Münch. Med. Woch.*, Aug. 15, 1889; *Neurologisches Centralblatt*, 1894, p. 506; *Deutsch. Zeitschr. f. Nervenheilk.*, 11 Band, 1 and 2 Heft, 1897.

Strzeminski, J.: "Troubles oculaires dans l'acromegalie" (three cases), *Archives d'Ophtalmologie*, Paris, February, 1897.

Surmont, H.: "Acromegalie a debut precoce," *N. Iconogr. de la Salpêtrière*, 1890, iii, 147.

Swanzy, H. R.: "Defective Vision and Other Ocular Derangements in Cornelius Magrath, the Irish Giant," Proc. Royal Irish Academy, Dec. 10, 1894.

Szymonowicz: "Experiments with hypophysis cerebri," *Pflüger's Archiv*, 1896.

Tamburini, A.: *Centralbl. f. Nervenheilk.*, 1894, v, 625; *Revue Neurologique*, November, 1897.

Tanzi, E.: *Riv. Clin.*, Milano, 1891, xxx, 533.

Taruffi: "Caso Della Macrosomia," *Annali Univers. di Med.*, 1879, t. 247; *Reale Acad. dell' Instituto di Bologna*, tome x, 1879, p. 63.

Thayer, W. S.: "Hypert. Pul. Osteo-arthropathy," *New York Medical Journal*, Jan. 11, 1896; *Philadelphia Medical Journal*, 1898, ii.

Thomas, J. Lynn: *British Medical Journal*, April 11, 1896; *ibid.*, June 1, 1895, p. 909.

Thomas: *Rev. Méd. de la Suisse Romande*, Geneve, 1893, xiii, 362.

Thomson, H. A.: "Description of an Acromegalic Skeleton," *Journal of Anatomy and Physiology*, London, 1889–90, xxiv, 475.

Thorne, Lesley Thorne: *British Medical Journal*, March 14, 1896; *Lancet*, 1896, i, 771.

Tikomiroff: "Vascular Changes in Acromegaly," *Presse Médicale*, No. 70, Aug. 26, 1896; *Rev. Neurologique*, 1896, p. 310.

Von Torday: "Congenital Giant Growth of the Extremities," *Jahrbuch f. Kinderheilkunde*, Bd. xliii, Heft 1, 1896.

Tresilian: "Myxœdema," *British Medical Journal*, March 24, 1888, p. 642.

Tschisch: *Deutsch. Petersb. Med. Woch.*, 1891.

Uhthoff: "Sehstörungen," etc., *Berliner Klin. Woch.*, 1897, p. 461.

Unna und Mendel: *Münch. Med. Woch.*, Dec. 3, 1895.

Unverricht: "Akromegalie und Trauma," *Münch. Med. Woch.*, 1895, xlii, 302, 329.

Valat: *Gaz. d. Hôp.*, 1893, lxvi, 1209.

Vassali: (See Sacchi.)

Verga: "Caso singolare de prosopeclasia," *Rendicont. del Reale Inst. de Scienze e Lettere*, April 28, 1864.

Verstraeten, C.: *Rev. de Méd.*, 1889, ix, 377.

Vinke, H. H.: *Medical Record*, 1896, ii, p. 779.

Virchow, R.: "Ein Fall und ein Skelet von Akromegalie," *Berl. Klin. Woch.*, 1889, xxvi, 81; *Deutsche Med. Woch.*, 1889, xv, 73; *Illustrated Medical News*, London, 1889, ii, 241.

Wadsworth: *Boston Medical and Surgical Journal*, Jan. 1, 1885.

Waldo, H.: *British Medical Journal*, 1890, i, 662.

Wells, H. Gideon: "The Thyroid Gland and its Congeners," *Journal of the American Medical Association*, 1897, p. 1009.

Whyte, J. M.: *Lancet*, 1893, i, 642.

Wilks: Clin. Soc. London, April 13, 1888.

Wolf, Kurt: *Beiträge zur Path. Anat.*, u. s. w., Zeigler, xiii, p. 629.

Woreester, W. L.: "Case of Acromegaly with Autopsy," *Boston Medical and Surgical Journal*, April 23, 1896.

Yamasaki, J.: *Kyoto Igakkwai Zashi*, 1893, No. 72.

Zeigler, Ernst. *Lehrbuch der Allgemeine Pathologie*, u. s. w., Baud i.

INDEX